AF342557

Statistical Process Control and Beyond

by

Richard R. Clements

Robert E. Krieger Publishing Company
Malabar, Florida
1988

Original Edition 1988

Printed and Published by
ROBERT E. KRIEGER PUBLISHING COMPANY, INC.
KRIEGER DRIVE
MALABAR, FLORIDA 32950

Library of Congress Cataloging-in-Publication Data

Clements, Richard R.
  Statistical process control and beyond.

  Includes index.
  1. Process control—Statistical methods.
I. Title.
TS156.8.C57 1988   660.2'81   87–4086
ISBN 0–89874–992–1

10 9 8 7 6 5 4 3 2

To Sally

TABLE OF CONTENTS

ILLUSTRATIONS

Figures

PREFACE

Several years ago, a man completed a life long dream by building a boat in his basement. Not until he had finished his dream boat did he realize that none of the doors out of the basement was large enough to allow the removal of the boat. Therefore, he promptly disassembled part of the house to create an opening large enough for the boat. The cost of this operation more than matched the expense of building the boat. Even though his dream had been realized, economically it had been folly.

Statistical Process Control has been almost universally handled the same way as the boat. This book was inspired by frustration. In most of the companies I have worked with, SPC was adopted as a new technique, but its full potential was ignored. The gains made because of information obtained from an SPC system were promptly offset by the failure to exploit the opportunities SPC created. Specifically, management rarely spent the time to encourage the strategic side of SPC, namely, the philosophy of continuous improvement. Today's production problems took priority over tomorrow's competitive position.

In the spring of 1982 I was working as a quality engineer and teaching statistical methods part-time. One day, the director of engineering at a local appliance manufacturer called me. He asked, "Have you ever heard of something called Statistical Process Control?" The engineers knew what a control chart was and had looked up the techniques of Statistical Quality Control (SQC). They were baffled by this name "SPC." Was it something new and not in their books?

I explained that SPC is the newest re-creation of the idea of SQC. Statistical techniques for controlling production processes have been around since the 1920s. Every couple of business cycles there is a new quality campaign launched by companies because of a perception that customers need better quality. Usually, these new methods fall by the wayside within a few years. Zero defects, SQC and others were past examples. What made SPC different was the idea that control charts would be kept by an operator who would make immediate corrections to the process. "Would this technique last?" they asked. "That would depend on the Japanese," I replied.

SPC was such a hot topic because study missions from the United States went to Japanese manufacturing plants during the 1970s and had seen the control charts in heavy use. The Americans were there to discover how an industrial nation, formerly held in low esteem, had been able to almost overwhelm the American markets in just under 10 years. The Japanese, it turned out, were introducing higher quality, lower cost products in a variety of markets. The study missions first concluded that the control charts and the strict following of SQC methods were the secret. Therefore, these representatives came back and told their respective companies that the way to compete with the Japanese was to use control charts and to require suppliers to do the same. The name adopted for this idea was "SPC."

Before 1983, I was lucky to discuss SPC with more than 100 people a year, and in 1984, I talked to and taught over 5,000. The rapid rise in demand came as Ford Motor Company and General Motors made SPC a major requirement of all their supply companies. At the time, over 7,000 companies were involved. Later, other large corporations would note the increase in automotive quality and productivity, and copy the requirement for SPC.

By 1985, the American delegations to Japan were finally getting the whole story about Japanese methods of production. Techniques such as Kanban, Just-in-Time, new vendor relations, and Taguchi's method of experimental design were coming to light. Each of these new technologies was developed and used with the assumption that an SPC system was in place to provide vital information about the capabilities of existing processes.

Now that I have worked with over 200 companies and thousands of employees, it has become clear that the true potential of SPC is being lost in a confusion of information about the method. People are focusing too much on the methods of SPC and not on the benefits of a well-planned and integrated system of continuous improvement. SPC must be seen and used as a complete system of management. SPC is not really a technique as much as it is an attitude—the idea that a company should examine itself and strive for improvement.

Rick Clements
Grand Rapids

ACKNOWLEDGMENTS

Appendix B. Information for the R & R studies was obtained from the *SPC Manual* by General Motors and from other company documents.

Appendix C, D, E and F. Information for compiling these tables was taken from Henry R. Neave's *Elementary Statistical Tables*, published by George Allen & Unwin Ltd., London, 1981. The information was cross-checked using Robert Mason's *Statistical Techniques in Business and Economics* and Robert K. Young and Donald J. Veldman's *Introductory Statistics for the Behavioral Sciences*.

CHAPTER 1
THE ESSENCE OF STATISTICAL PROCESS CONTROL (SPC)

Statistical Process Control (SPC) is a paradox. It can be thought of as an individual method of monitoring production processes or as a complete system of management. In either capacity, its results for a company can range from highly profitable to extremely harmful. It is all a matter of how the SPC system is introduced and used within a company. For maximum benefits, the SPC system must be transparent—in other words, it must appear to be the logical way to do things. This requires the replacement of some traditional assumptions about manufacturing and management.

FORMS OF SPC

SPC comes in two forms. The first is the familiar tactical form—the day-to-day use of control charts and capability studies. The second form is strategic which uses SPC as a gateway technology. SPC is the first step in the implementation of a wide variety of manufacturing technologies. The knowledge gained by a complete SPC system helps to advance the competitive position of a company. For example, the charts and studies of manufacturing processes may find that through minor redesigns of the product, a little automation, and a faster turnover of inventory, a company could increase quality and reduce production costs. Examples of these connections between the tactical SPC and the strategic opportunities they create will be the backbone of this book.

The principal philosophy of SPC is the idea of continuous improvement. However, the actual technique of SPC is best interpreted as "stable, predictable, and capable." A process is studied by drawing regular samples of its production and examining critical characteristics of the product. For example, if a machine is supposed to cut wires six inches long, then the average length of wires being produced should be six inches. If an operator checked every day and always found the process produced an average wire length of six inches, the process would be called stable. Naturally, the actual lengths of each wire will vary slightly. However, if the variation is consistent day after day, the operator will find the amount of variation to be reliable—or in other words, predictable. Finally, the specification may call for wires that are six inches, plus or minus two hundredths of an inch. If the historic estimated variation the operator has observed for wires is less than this specification, the process is capable of meeting specifications. In short, the process is stable, predictable, and capable.

To accomplish the situation above assumes the operator stops at regular intervals and checks production. The information the operator obtains is recorded, and some calculations are made. A prediction of normal variations will be made and noted. When a process fails to perform as predicted, corrective actions are taken on the assumption that something has changed in the process. The action is noted by the operator and the process is checked again to confirm the correction. This procedure is called "maintaining statistical control." The next two chapters will show how this is done with control charts and capability studies. The important aspect of this method to remember is that the operator is controlling the processes, based on real-time information, and established goals. This procedure replaces the method of running a process for maximum production and sorting out defective goods later. Its full implications will also be explored in later chapters.

HOW SPC BECOMES TRANSPARENT

SPC is like any statistical method of analysis in that it can not stand alone; statistical analysis means nothing by itself. To have any benefit it has to be applied to making decisions in the real world. The same is true of SPC. Some people see SPC as a collection of control charts in the production area, and some SPC systems are just that. Unfortunately, such systems are worthless unless the charts are tied into the decision-making processes of the company. The rest of this book will describe how the complete SPC system is created when the control charts become the central source of information for continuous improvement in all departments.

IMPLEMENTATION OF SPC

To implement SPC, its methods and philosophies must be carefully connected to every aspect of running a company. The procedure begins by focusing each department's attention on the continuous improvement of the manufacturing process. (Later this will be spread to all processes in the company.) SPC is the replacement of older management techniques such as concentrating on quarterly profits, rigidly following a production schedule, observing the concept of quality control, and telling individuals what to do. By questioning a lot of previously well-guarded assumptions, it promotes teamwork and studies the processes and methods of production and competition.

In short, SPC is best implemented when it is tied into existing techniques as the idea of continuous improvement, and when SPC is used as the introduction to new technologies. As later chapters will show, SPC is the logical first step for techniques such as Just-in-Time deliveries, Zero Inventories, automation, and Computer Aided Manufacturing. At the same time SPC should be part of the production control process. The operators of an SPC system are the operators of the manufacturing processes. The inspection function is eventually converted into a quality assurance group that audits and supports SPC. The engineers, managers, supervisors, and technicians in the plant benefit from the information produced by the SPC system.

THE SPIRIT OF SPC: CREATIVITY AND INNOVATION

SPC is a dynamic system. In a complete SPC system, work teams are formed to use the information about process problems. These teams engage in active problem-solving activities to continually improve the efficiency and effectiveness of manufacturing processes. Teams rely on leadership to continue their existence, but creativity and innovation are the main tools used to find solutions to production problems. The skills, knowledge, and experiences of the production workers are tapped as a vital resource for increasing the competitive position of the company which is a significant change in the traditional view of the role of the production work force.

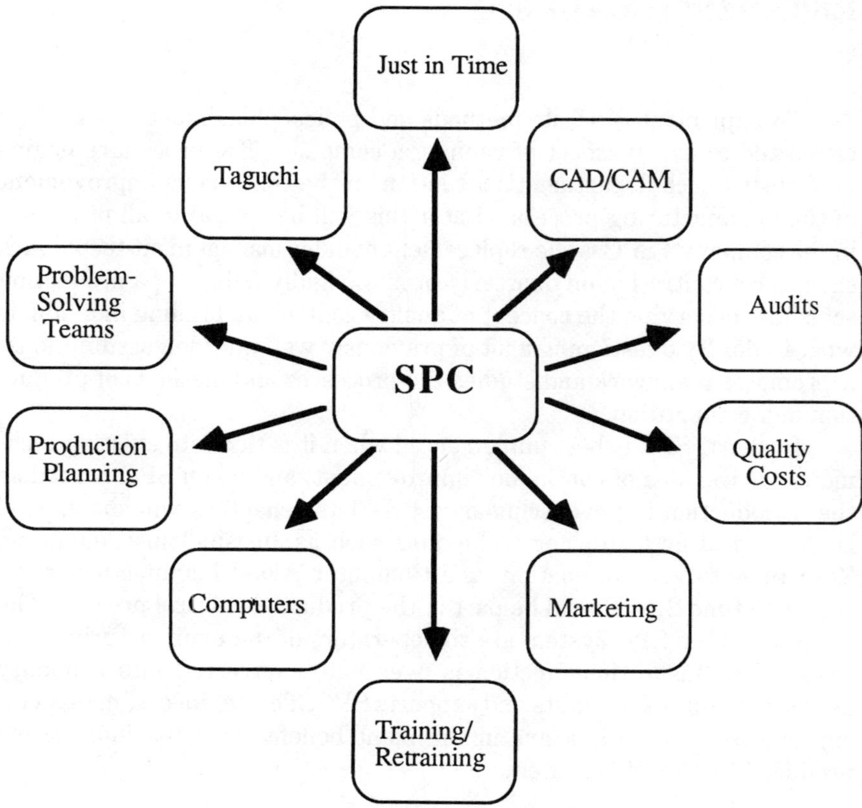

Figure 1.1 The connections between SPC and other modern manufacturing technologies. SPC proves to be an excellent starting point in the implementation of new methods.

SPC AS A SOCIAL REVOLUTION

The final chapter of this book is a review of the human impact of SPC. Its social effect is only now being felt by companies. SPC changes the social order within a company. First, it encourages employees to work together as teams, sharing common goals and problems. Second, to implement SPC, a massive training program is required that exposes most employees to classrooms for the first time in many years. Although problems such as illiteracy and math anxiety will be encountered, one benefit has been duly noted—some people enjoy learning. Companies are discovering that improving the employees' skills makes them feel more a part of the company. Third, and most important, is giving primary control of the production process to the operators who monitor and correct the process on their own initiative. Managers need to do less managing and more supporting of the operators and their work teams. The final effect of these three major changes can only be guessed, but change in the way the work place functions is certain.

THE BOTTOM LINE: COSTS

The pursuit of higher quality and a more competitive position for a company is a secondary effect of a greater motivation i.e., costs. The recent surge in the use of SPC and other related manufacturing technologies is not the result of companies suddenly gaining an interest in greater efficiency and customer satisfaction. Instead, it is the direct result of the invasion of markets by overseas competition. Although SPC can improve the quality of production, its main goal is to reduce costs. Only by reducing costs through continuous improvement of processes and products can a modern company hope to survive. Thus, SPC can also be seen as an economic tool. The information gathered by any SPC system must include the economic impact of process capabilities and product quality. Only then can management intelligently direct the proper corrective actions.

WHY MANAGEMENT MUST CHANGE

Today's customer demands a higher level of quality in the products and services in the marketplace. There are several reasons why this is so. For example, in the 1960s and 1970s there was a rise in consumer concerns over the safety of many products. The emergence of the Japanese products in the marketplace introduced serious competition, and a consumer reaction against the "throw away" products is another possibility. However, the largest force is economics. The recent recessions and the effects of the oil embargo of 1973–1974 made cash a more precious commodity. Consumers wanted more value in products and services to offset the lower level of available cash. This increase in value is interpreted as an increase in quality—specifically, a product that is more durable, attractive, and reliable. For a service, this means people who are prompt, polite, and provide timely, accurate information.

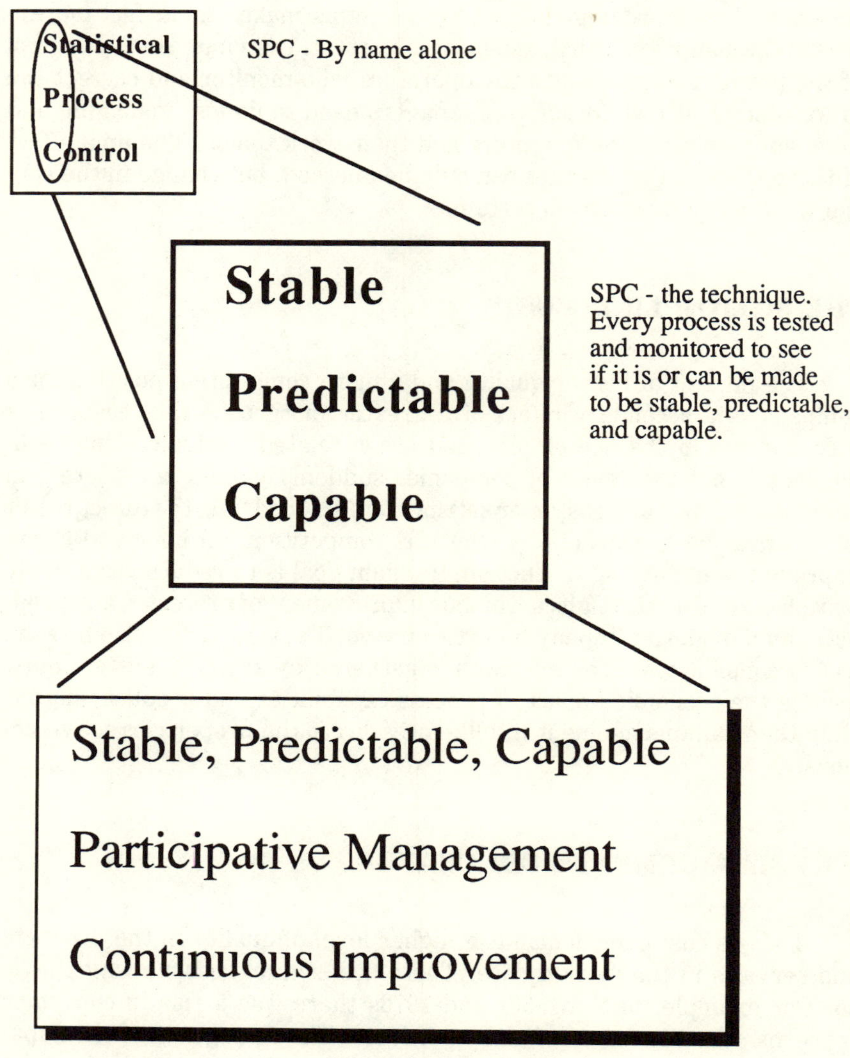

Figure 1.2 The three faces of SPC—name, technique, and complete
    system.

Every authority in quality from Feigenbaum to Crosby to Deming to Juran has declared a new economic age: an age where a world economy has replaced nationally based industries; an age where the consumers drive the level of quality a company produces; a time where companies must cut costs and raise quality to stay competitive. In short, a time when the only chance for a company to survive is by continually improving its methods, people, products, and services.

SUMMARY

SPC is neither an independent technique, nor a new form of quality assurance. Instead, it is a new management style which is based on the need to know how good the manufacturing processes are and what problems haunt the production personnel. Management under SPC actively seeks out both problems and solutions by forming cooperative work teams that monitor the day-to-day activities of the company. The specific information gathered by these groups is used for their own problem-solving efforts. General summaries of the information go to management so it can direct the strategic activities of the company.

The following chapters will detail the complete SPC system. This will include discussion about how it is implemented, how it will benefit the company, and what other technologies will tie into the system.

CHAPTER 2
CONTROL CHART FUNDAMENTALS

SPC is a system of using knowledge to compete. A company can increase its chances of expanding markets by learning as much as possible about its own processes. The technique begins by studying the normal pattern of behavior a process exhibits and the tool for studying the behavior of a process is the control chart. The control chart is a mechanism for directly monitoring the stability, predictability, and capability of a process and has three main components. The first is the process capability study that is a short term examination of the quality of products manufactured by the process in question. The second part is the actual control chart and its ability to monitor a process in real time. The third part is the process log for recording the day-to-day activities related to operating the process. With these three parts working together, a company will create enough information to make continuous improvement possible.

ASSUMPTIONS IN PLACE
BEFORE SPC AND CONTROL CHARTS EXIST

The introduction of control charts implies the displacement of an existing system of control for a process. No process operates in a void. The typical system of control before SPC is one related to the production schedule. Operators often refer to this system affectionately as "quantity control." The ideal is to meet the production schedule every day with few breakdowns or shortages of materials. If a problem does occur, then the operator calls a die setter, materials control person, or somebody else to correct the situation.

An SPC system of manufacturing is vastly different from this form of "quantity control." The emphasis on the production schedule means that other activities unrelated to direct productivity are discouraged. The ideas of continuous improvement, stopping to monitor a process, and operators meeting to discuss better production methods, are not allowed. Little is learned about how the process works, and advanced planning rarely goes beyond the next 24 hours. SPC, and especially control charts, constitute a sweeping reformation of the manufacturing process. Chapter 4 is a detailed description of the complete system.

EXPLANATION OF A CONTROL CHART

A control chart is a tool that enables the operator to directly monitor the quality of production from a given process. The oil light in a car works by the same principle. Without an oil light, the operator of a car would have to rely on periodic inspection to ensure that the oil levels are correct in the car. Inspection of the oil system while traveling down the highway would be impossible. The first indication of a problem would come after low levels of oil caused damaging friction inside the engine. Then corrective action would take place only after damage had occurred.

An oil light, on the other hand, is connected to a sensor that continuously monitors the oil pressure in the engine. When it is low, the light activates, the driver normally pulls to the side of the road, and checks the oil system. If all works well, the car's operator is alerted to a potential problem before damage occurs. Prevention replaces detection.

While the oil light is off, most people keep driving. When it flashes on, most react by stopping. However, when the light comes on, it is only indicating the possibility that a problem exists. Sometimes everything is fine when the light comes on, but, it is still a good idea to pull over and check the oil system. Maybe the light is faulty and needs replacement, or maybe the oil system does have a problem. In either situation, reaction is still appropriate.

The control chart works very much like the oil light. A sample of the production is checked regularly. Any deviations from normal behavior of the process are noted by the operator and the operator reacts. The control chart monitors the stability and predictability of the process in real time. The operator is able to use this information to take preventive actions, rather than waiting for a problem to justify corrections to the process. Just like the oil light, a significant deviation on the control chart does not mean that a problem exists, only that there is potential for a problem.

Basically, there are two types of charts. The first family of charts is called variable data control. These are the control charts for situations where data will be gathered using a variable scale of measurement. Examples of such scales include length, width, depth, diameter, temperature, and pressure. The other family of control charts is for data gathered about attributes. Attributes are the characteristics of a product or process

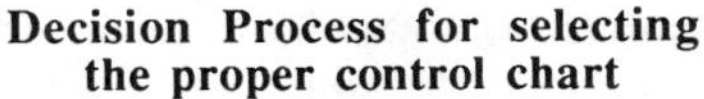

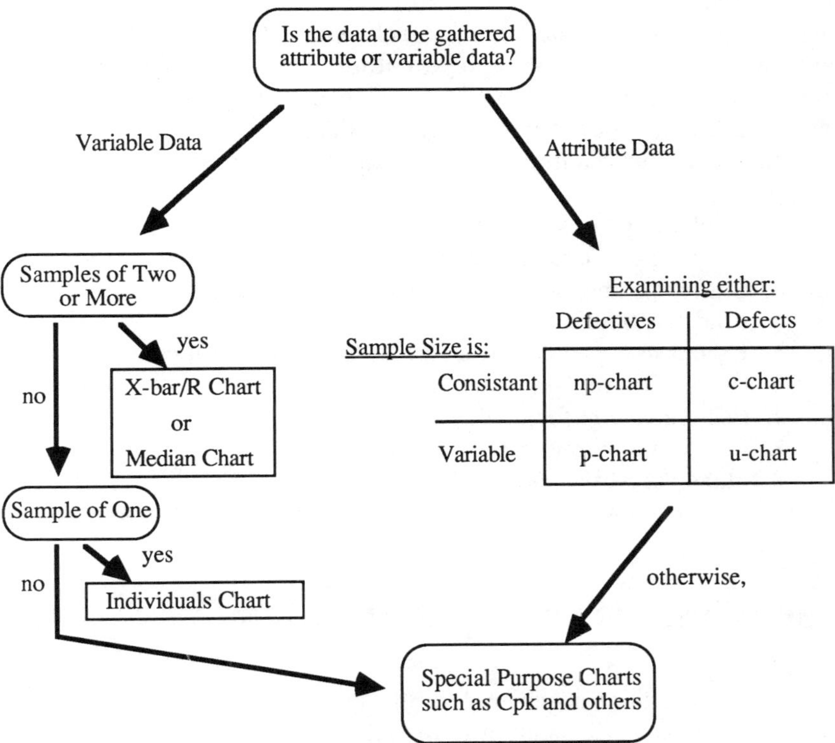

Figure 2.1 The decision tree for selecting the correct control chart.

and represent an either/or situation. Examples include defects, scrap, and rework. The next chapter will address the issue of attributes more fully. The following illustration will explain the functions of a variable control chart in detail.

CONTROL CHART IMPLEMENTATION

The heart of an SPC system is its control charts which help the operators to gain some sort of control over the normal behavior of the process. In addition, these control charts will be the focus point for the integration of the SPC system into the operations of all other areas of the company. In the end, control charts will be the key competitive tool for the company. Below, a mythical furniture factory will illustrate the creation and use of a variety of control charts. All of the steps toward implementation of this chart apply to other forms of control charts. The only exception is that the formulas used to create the control limits will change according to the type of chart used.

The steps towards implementing a control chart are:

1. Meet the process.
2. Select the critical characteristic.
3. Study specifications and set-up.
4. Find the baseline.
5. Estimate expected variation for the creation of control limits.
6. Create the control chart.
7. Post the chart for 30 days.
8. Perform a second capability study.
9. Make corrections/updates.
10. Use the chart for the continuous improvement of the process.

CASE STUDY: THE XERON FURNITURE COMPANY

In the illustration of the implementation of a control chart, a mythical furniture company will attempt to install a variables control chart. More specifically, it will use an Average and Range chart, also called the X-bar/R chart (X-bar and R are statistical shorthand for average and range). The piece of furniture in question is a bookcase. A meeting of operators, and personnel from engineering, sales, and quality control reveals that the largest customer complaint is that the shelves tend to warp or come loose. Records from the shop floor also indicate that a lot of scrap and rework are associated with these shelves.

Step 1: Meet the Process

Whether it be wire cutting, metal stamping, welding, injection molding, bill of material creation, or any other process, the first step is always to look at the process. An overview of the process and its environment is necessary. A process can include machines, people, materials, methods, environment, and paperwork. The use of all five senses is recommended because this knowledge of the situation could be invaluable later on. For example, if a work area has a lot of greasy parts, a company would take steps to protect the control charts with a plastic covers.

At this point the group from the above illustration would head out onto the shop floor to inspect the location where the wood is delivered to the plant. Their first task would be to explore the process of building bookcases by following the wood through the plant until it is shipped as a finished product. By doing this, they learn that the pine boards are first planed and then sent to a cutting area. In the cutting area the length of the shelves is determined. They examine the cutting procedure more closely and, one team member notes that a table saw does the cutting and that a guide holds each board to what is assumed to be the proper length. Other team members also observe the abundance of sawdust in the area. In fact, it seems to gather in places such as at the edge of the adjustable guide and that the saw seems to flex a little if a piece of wood is forced through. These are all important observations for use later on.

In the next stage of production the components are assembled. Also in this area shelves that are too short to fit are thrown onto a scrap pile. When a team member asks about pieces that are too long, the foreman points to a handsaw in the corner and indicates that it is used to trim excess from shelves. The scrap pile and the rework with the handsaw are two factors that add to production costs. Eliminating them by making the boards more consistent would result in a considerable cost reduction which the team also notes for later use.

After assembly, the bookcases are cleaned with an air hose located in a special booth that vents the sawdust outside. Then the bookcases are moved to a vented room for staining and varnishing. Staining is done by hand, and the lacquer is applied by compressed air gun. The bookcases then cure for 24 hours in a special dust-free waiting area. After curing, they are packaged and sent to shipping which confirms the contents and labels them for shipment. The team stops exploring the process at this point.

The team has explored the entire process as far as possible. Thoroughness is necessary since later problems and their effects could spread into areas never suspected when they first began creating the chart. In most applications of control charts, the team would also review the suppliers and consider the customer's needs.

Step 2: Select the Critical Characteristic

SPC is not intended for monitoring all the characteristics of a product. It is based on the idea that at each stage of production one characteristic will be critical to the product. It is called the "critical characteristic," or is sometimes referred to as the "control item."

Selecting a critical characteristic involves looking at the dimensions or attributes important at each stage of production. A list of these can be drawn up from a meeting with the operators, the engineers, purchasing, sales, and quality control personnel. Each stage of the production process is evaluated for a critical characteristic using the following three-level priority system:

1. **Customer Requires It.** In this case a control chart is mandatory and should follow the customer requirements or suggestions. These are usually found on the blueprints or within the customer's purchase orders.
2. **Important to Internal Operations.** If a characteristic of a part could be monitored to help reduce internal problems with production, then a control chart should be used. For example, if all subassemblies are bolted together in four places, the important factor is in getting the proper hole locations drilled before assembly so that all subassemblies will bolt together properly on the first try. The customer cares only about the subassemblies being bolted together properly. However, the company is intensely interested in the first-time success rate of the operation. The more first-time successes, the less rework costs involved.
3. **Item of Interest.** All customer requirements and items important to the company should be exhausted before selecting an item of interest. For example, a receiving area may have no problems with the flow of inventory, but someone may monitor time to find out if hidden potential for improvement is present.

In the Xeron case, many dimensions were important to the construction of a bookcase. The function of the team was to sort out the ones most important for the successful operation of the final product.

A further discussion with an engineer revealed two primary causes for the warping reported by the customers. The first occurs when a shelf is a little longer than it should be. When assembled, it is placed under stress across its length and wood tends to warp under these conditions. The other possible cause cited is the wood not being properly sealed by the stain and varnish so that moisture then can enter the fibers of the wood and cause warping. Remembering the staining area, it occurred to one team member that inconsistent shelf lengths would mean gaps in the assembled bookcase—perhaps these gaps let the moisture in. Having an overview of the situation, the team agrees that the length of the shelves seems to be the major cause of scrap and rework. The final decision of the group is to select the shelf length as the critical characteristic in the cutting process.

The team also designates other critical characteristics such as surface coverage in the finishing area. Chances are that more than one of these critical characteristics causes the warping problem. Chapter 3 will explore the monitoring of these other critical characteristics.

Step 3: Study Specifications and Set-up

During set-up operations, someone should take careful notes of the procedure used and any problems encountered. This will help in future setups by providing tips for faster and more successful preparations. The records of each setup for a particular job are combined to note significant differences. Perhaps a particular method increases the capability of the process, and without documentation, there would be no way to know this.

Obtaining the blueprint for the bookcase, Xeron's SPC coordinator heads back to the cutting area with the foreman and the operator and watches the setup of the table saw for cutting shelves. The specifications call for a four-foot shelf, plus or minus an eighth of an inch. The operator sets up the saw by first loosening the guide and then measuring the distance between the inside edge of the saw blade and the guide. The distance is set to four feet and the guide is tightened down. As a test, a scrap piece of pine is passed through the saw and measured. If it is within an eighth of an inch of four feet, then the process proceeds. The measurements are all made with the operators' tape measures. Each operator has a tape measure which should be noted for later discussion. The first job is to see what the process is doing under day-to-day conditions.

Step 4: Find the Baseline

Once the critical characteristic is selected and evaluated, the related process can be studied. The intention is to learn how a particular process works when it is protected from as many variable factors as possible. First the specifications, seen as guidelines, must be studied. The effort should be focused on comparing results to specifications by testing the process as if the specifications were unknown. The specifications can be used later to evaluate the process potential of the machine.

The Process Potential Study. Process potential can be established with a 30-part capability run for variable data, or a 100 piece run for attribute data, after a few pieces have been run on the process to confirm that all settings are properly placed. No alteration of machine settings is allowed while the study pieces are being made. If possible, only single sources of material and the same operators are used. The operator is instructed to cut 30 shelves without any adjustments to the process, meaning the guide can not be adjusted, the blade can not be changed, and the sawdust can not be removed.

All pieces should be evaluated for their critical characteristic. The average count, proportion, or measurement establishes a baseline. A baseline is a rough guess as to the average quality of a typical part. If 30 pieces are made and found to have an average overall length of 4 feet and 1/64 inch feet then, the 4 feet and 1/64 inch average would be the baseline of the process. Over time the average size of parts sampled from this process should be close to 4 feet and 1/64 inch. However, if the specifications call for an optimal overall length of 4.00 feet, the process would have to be adjusted to the exact specified standards so that the process baseline is "on target" with the specifications.

After cutting 30 pieces, the operator stacks them for measurement. Since the SPC coordinator already suspects that tape measures are not the best way to measure, a steel ruler with 1/64 of an inch increments will be used. A metal square is held at one end of the ruler to further ensure the accuracy of the measurement. In Chapter 9 there will be a strong recommendation that any system of measurement should be thoroughly tested before being used in the production area.

The following 30 readings are obtained. Each is expressed as to how many 64th's of an inch each one is above or below four feet.

A quick calculation reveals that the average board is indeed four feet long. Had this not been the case, corrective action would be necessary to adjust the saw to the proper length. If the boards were 1/16 of an inch short on the average then, the guide could be moved 1/16 of an inch away from the saw, even though the average is within specification. The advantage of making the change is to increase the predictability of the process. After all, the next process, assembly, is based on the assumption that the lengths of the shelves are four feet.

Step 5: Estimate Expected Variation for the Creation of Control Limits

Once the baseline has been established, the amount of expected variation is estimated by calculating the standard deviation for the raw data. The first step is to create a histogram from the 30 pieces of data. It is critical to see if the data seems to be forming a normal distribution. The curve of normal distribution, also called a bell curve, is interesting, but rarely explained statistically. To see why it is called the "normal curve", just put about 100 pounds of beach sand in a bag and hang it above the

The bookcase work team conducts a process potential study by measuring
30 randomly selected shelves. They use a steel ruler to measure each
shelf to the 64th of an inch. Each reading is recorded as the number of 64ths
above or below the optimal length of four feet.

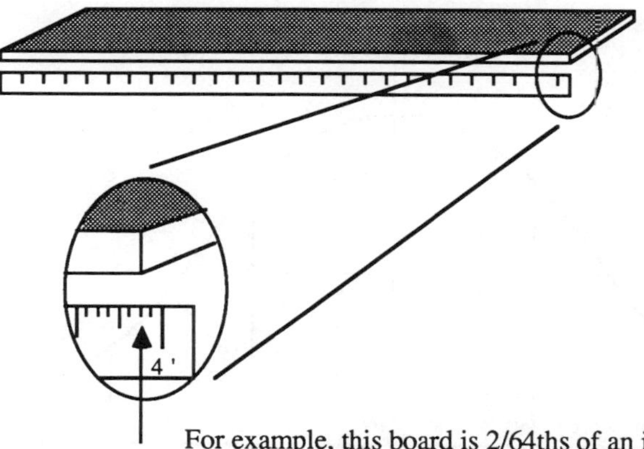

For example, this board is 2/64ths of an inch short of
four feet. Therefore, it is recorded as " -2 ."

The resulting figures were obtained for the 30 pieces measured.

0	4	2
2	0	2
4	-2	-2
-2	4	0
4	0	8
2	2	-2
0	6	-4
-6	-2	0
-8	-4	-4
2	-6	0

Table 2.1 The collection of data for a thirty-piece process potential study

Curve of Normal Distribution

In this example, the average is equal to 100 and the standard deviation
is equal to 5.

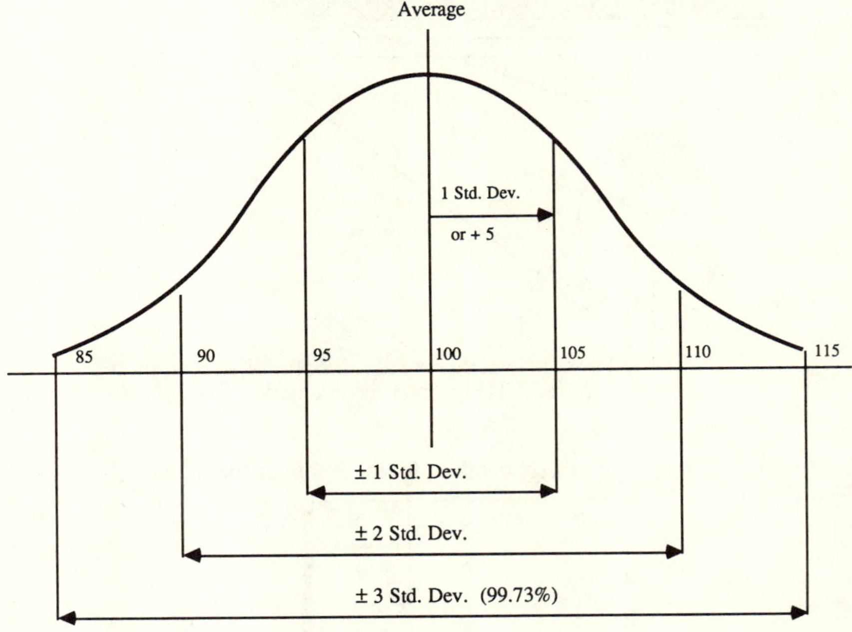

The curve predicts that 99.73% of the production will fall between 85 and 105.
This is known as the expected amount of variation.

Figure 2.2 The curve of normal distribution and approximate area under
the curve.

ground. Punch a hole in the bag and the sand will pile up into a bell-like
shape because this is the way sand will normally distribute itself.

When the distribution is pushed to one side like the pile of sand being
blown in one direction, the distribution is "skewed." When observing a
sand dune, the effect of a single identifiable force, the wind, is apparent.
The same holds true for variations in the process. If the distribution is
something other than normal, then it is probable that a single identifiable
factor is affecting the process. In this example, the distribution is somewhat normal. Only when the distribution appears normal, can the standard deviation be calculated.

Standard deviation is a universal index of variation. That means it

The Standard Deviation Formula:

$$\sqrt{\dfrac{\Sigma\, x^2 - \dfrac{(\Sigma x)^2}{n}}{n - 1}}$$

$\Sigma\, x$ = the sum of all readings

$\Sigma\, x^2$ = the sum of all readings squared

n = the number of readings

Calculating a Standard Deviation

Find the Standard Deviation of the following numbers: 4,2,5,4,5,2,6

Step 1: Place all numbers in a column and create a second column of squares.

X	X^2
4	16
2	4
5	25
4	16
5	25
2	4
6	36
$\Sigma x = 28$	$\Sigma x^2 = 126$

Step 2: Calculate.

$$\sqrt{\dfrac{126 - \dfrac{28^2}{7}}{7 - 1}}$$

or, $\sqrt{2.3\overline{3}} \quad = \quad 1.53$

Figure 2.3a The standard deviation formula and an example of its calculation.

Calculating an Average:

$$\bar{x} = \frac{\sum x}{n}$$

where, $\bar{x}$ = The mean average

$\sum x$ = The sum of all readings

n = The number of readings

Example of calculating an average using the data from Table 2 - 1.

0	4	2
2	0	2
4	-2	-2
-2	4	0
4	0	8
2	2	-2
0	6	-4
-6	-2	0
-8	-4	-4
2	-6	0

Sample Size (n) is 30 pieces.

Total ($\sum x$) = 0

$$\text{Average } (\bar{x}) = \frac{\sum x}{n} = \frac{0}{30} = 0.0$$

Figure 2.3b Calculation of an average (mean).

expresses standard amounts of variation regardless of the units of measurement. Standard deviation also estimates the variation to expect in future boards that will be cut. The following examples illustrate this.

Using the 30 measurements taken we obtain a standard deviation of 3.68. When the model of the curve of normal distribution is referred to, 99.73% of the area under the curve is accounted for by a zone three standard deviations from each side of the average. This means that six times the standard deviation, produces an estimate of the range of variation for 99.73% of the parts. In many industrial applications, 99.73% is assumed to be close enough to 100% to serve as the estimate of total variation. In the example, six times the standard deviation of 3.68 equals 22.08, or $^{22}/_{64}$th of an inch. That is the distance from the smallest board that will be cut to the largest that will be cut.

By comparing these results to the original specifications, it can be determined whether the variation expected in the process could be contained by the specification limits. This can be seen with a picture or calculated mathematically. One such calculation is called the Cr Index, or the capability ratio and it expresses how much of the tolerance range is consumed by the expected variation of the process.

A result of .75 or less is ideal, meaning that the process is only allowed to use up to 75% of the tolerance range. A result of 1.00 indicates that 100% of the tolerance range could be used up by the process variation. Assuming the average board is cut to four feet, the process can produce board lengths anywhere within the entire tolerance range. Results higher than 1.00 predict that scrap work will be unavoidable, because the process variation is larger than the tolerance spread.

In the Xeron example, the tolerance range was plus or minus $^{8}/_{64}$ths of an inch. This is a total tolerance range of $^{16}/_{64}$ths of an inch. The expected process variation was $^{22}/_{64}$ths, which results in a Cr index of 1.38. Clearly, the process must be improved before it is certified stable, predictable, and capable. Corrective actions are taken until the Cr index becomes more acceptable to management. Then the control chart can be created for use with this process.

Control Limits. Anytime the board cutters sample, perhaps, five boards, an average length near the expected four feet is produced. Yet, how far can one of these averages be off before there should be concern? The answer is in the control limits. Control limits are the boundaries of chance. Anytime five boards are randomly selected, there is a chance of drawing five boards that are all shorter or longer than the average four foot length, meaning the observed average from five boards will usually be different from the true average of the process. Control limits show how far the observed average for a given sample size (five in this case)

Raw data from Table 2 - 1.

0	4	2
2	0	2
4	-2	-2
-2	4	0
4	0	8
2	2	-2
0	6	-4
-6	-2	0
-8	-4	-4
2	-6	0

Frequency Distribution

Dimension (in 64ths)	Frequency
-8	1
-6	2
-4	3
-2	5
0	7
2	6
4	4
6	1
8	1

The information in the frequency distribution is used to create a histogram to check for a normal distribution.

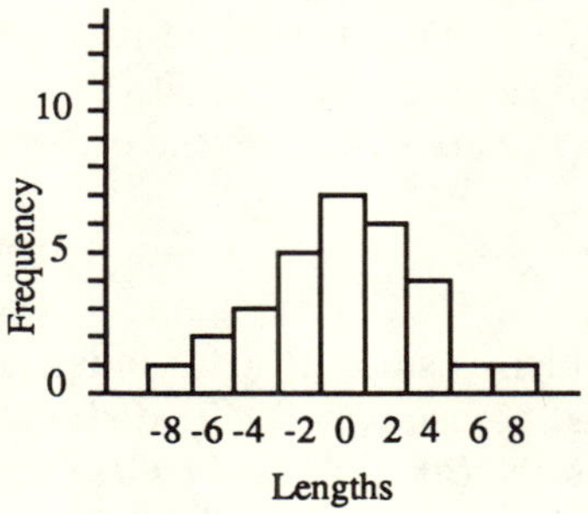

The raw data is used to calculate a standard deviation.

$$\sum x = 0 \qquad \sum x^2 = 392 \qquad n = 30$$

$$\sqrt{\frac{\sum x^2 - \frac{(\sum x)^2}{n}}{n - 1}}$$

or,

$$\sqrt{\frac{392 - \frac{0}{30}}{29}}$$

thus, Standard Dev. = 3.67658

Figure 2.4 The conversion of raw data into a frequency distribution, then into a histogram (to check for the normal distribution), and finally into a standard deviation. These steps produce a picture of the variation occuring in the process.

For all examples, assume a tolerance of 4.00 feet $\pm$ 8/64 ths of an inch.
An average of 4.00 feet and a standard deviation of 3.67 (64 ths).

The Cr Index of Capability

The Cr Index is the measurement of how much of the tolerance spread is being "burned up" by process variation. The ideal process will have a Cr index of less than 0.75 (or 75%).

Use either a calculator or the $R \div d_2$ method to estimate the Standard deviation. Then calculate...

$$\frac{6 \times \text{Standard Dev.}}{\text{Tolerance Spread}} = \frac{6 \times 3.67}{16} = 1.38 = \text{Cr Index}$$

The Cp Index of Capability

The Cp Index is the same as the Cr, but you flip the equation over.

$$\frac{1}{\text{Cr}} = \frac{1}{1.38} = 0.725$$

This number is 1.33 or higher for the ideal process.

Figure 2.5 The calculation of Cr and Cp capability ratios.

can deviate by chance from a true process average. Dotted lines on a control chart define the point where an observed average is no longer the result of chance. Beyond these control limits is the area known as "out-of-statistical-control".

Step 6: Create the Control Chart

Once a process demonstrates the potential for exceeding specifications, the next step is to create a control chart to see if the process is stable, predictable, and capable over time. At regular intervals, samples

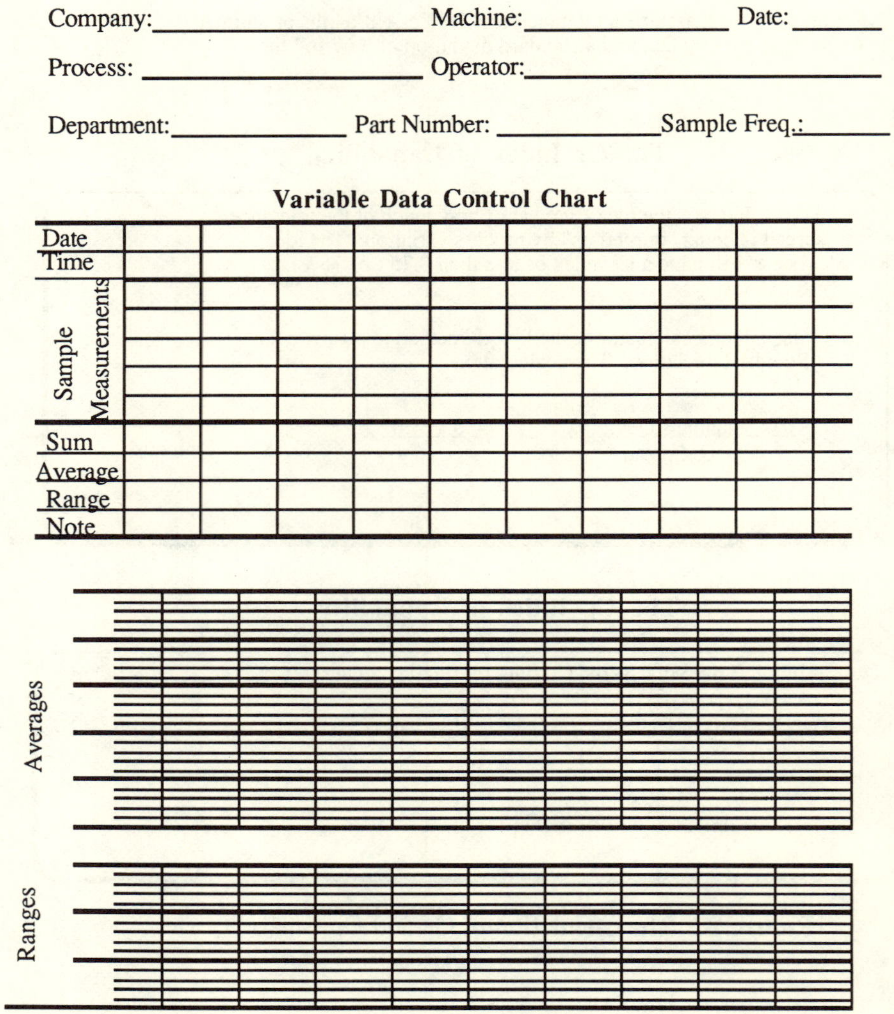

Company:_________________________ Machine:_______________ Date:_______

Process: ___________________ Operator:_______________________

Department:_____________ Part Number: ____________Sample Freq.:_______

Variable Data Control Chart

Figure 2.6 Enlarged portion of a variable data control chart. This chart will be filled out as an average and range chart.

of production are taken and the average measurement or count is recorded. Twenty to 25 of these samples at regular intervals must be taken before control limits can be established.

Note on sampling frequency:
A question frequently asked is how often to sample a process. The answer depends on the rate of production, the importance of the

part, and the difficulty in measuring. The following are rules of thumb and should be used only as guidelines. Processes that produce over 1,000 parts per hour should be sampled about every half an hour, if measurement is relatively easy. Production between 100 and 1000 per hour should be checked every hour. Fewer than 100 parts per hour should be checked at least once a shift. Critical components, such a rare alloy castings that costs thousands of dollars to produce, should be checked more frequently.

The information from the 20 to 25 samplings is used to calculate a new baseline and control limits. The baseline should be a continuous line across the chart, with the control limits shown as heavy dashes above and below the baseline. The actual values of the baseline and control limits should be noted somewhere on the chart. Also noted, should be the name of the part, its critical characteristic, how to measure the part, time and dates of samplings, and the fact that this is the first chart in a series. When possible, the process is adjusted so the baseline is at the optimal specification.

For each group of five parts, an average and a range were calculated. The averages and ranges are plotted on the graph area above the data columns. Next, the 25 means are averaged resulting in an X-double bar or "process average". Then, the 25 ranges are averaged together to calculate the R-bar or "average range". Both the process average and the average range are drawn as horizontal lines on the charting paper.

Control limits are calculated using the X-bar/R control limit formulas below. The resulting control limits indicate how far averages and ranges can deviate from their respective baselines by chance. Since this is the preliminary chart for the establishment of control limits, points beyond the control limits are not yet the primary concern. The first concern is to talk to the operator and to examine the chart to see if there are flaws in the SPC system. In this case, the operator wants to know where to record that the saw blade was changed. It is agreed that the time column corresponding to the blade change will be marked with an asterisk.

As a double check of the control chart, a histogram is created using the reading taken over the 25-hour period, making a total of 125 readings in this example. The histogram should be roughly a normal distribution. Computer software can help determine how close to normal the histogram is. In the example, the curve is still rather flat, just as in the first capability study which will be kept in mind as a control chart is prepared for the cutting area.

Some SPC coordinators save time by placing a blank control chart at the process of interest and instructing the operator to gather the initial

Time	8	9	10	11	12	1	2	3	4	5
1	0	4	2	2	0	2	4	−2	−2	−2
2	4	0	4	0	8	2	2	−2	0	6
3	−4	−6	−2	0	−8	−4	−4	2	−6	0
4	4	0	8	2	2	−2	0	6	−4	0
5	2	−6	0	−8	−4	−4	−6	−2	0	2
Sum	6	−8	12	−4	−2	−6	−4	2	−12	6
Average	1.2	−1.6	2.4	−0.8	−0.4	−1.2	−0.8	0.4	−2.4	1.2
Range	8	12	10	10	16	6	10	8	6	8

The data is used to calculate the average of the ten averages (the process average)
and the average range (R-bar).

$$\bar{\bar{X}} = \frac{1.2 + (-1.6) + 2.4 + (-0.8) + (-0.4) + (-1.2) + (-0.8) + 0.4 + (-2.4) + 1.2}{10} = 0.4$$

$$\bar{R} = \frac{8 + 12 + 10 + 10 + 16 + 6 + 10 + 8 + 6 + 8}{10} = 9.4$$

The process average and the average range are combined
with control chart factors to create the control limits.
(see table 2 - 2 for control factors)

$$\text{Control limits} = \bar{\bar{X}} \pm (A_2 \times \bar{R})$$
$$\text{or, } +0.4 \pm (.577 \times 9.4)$$
$$\text{Upper Contol Limit (UCL)} = 5.8$$
$$\text{Lower Control Limit (LCL)} = -5.0$$

$$\text{Range UCL} = D_4 \times \bar{R} = 2.114 \times 9.4 = 19.87$$

$$\text{LCL} = D_3 \times \bar{R} = 0 \times 9.4 = 0$$

Figure 2.7 The calculation of control limits using data from the bookcase
example.

20 to 25 samplings. This helps orient the operator to the type of data to
be gathered once the chart is implemented.

The Xeron group began by placing a blank sheet of X-bar/R charting
paper next to the saw in the cutting area. The operators were given eight
hours of training in the basics of SPC that emphasized the advantages of
monitoring a process and how to fill in a sheet of control chart paper.
After every hour of cutting, an operator would select five randomly chosen
boards from the hour's production. These boards would be measured with

Control Chart Factors

n	A_2	d_2	D_3	D_4
2	1.880	1.128	-	3.267
3	1.023	1.693	-	2.574
4	0.729	2.059	-	2.282
5	0.577	2.326	-	2.114
6	0.483	2.534	-	2.004
7	0.419	2.704	0.076	1.924
8	0.373	2.847	0.136	1.864
9	0.337	2.970	0.184	1.816
10	0.308	3.078	0.223	1.777

To calculate the control limits for the averages, use the following two formulas:

$$\text{Upper Control Limit } (\overline{X}_{UCL}) = \overline{\overline{X}} + (A_2 \times \overline{R})$$

$$\text{Lower Control Limit } (\overline{X}_{LCL}) = \overline{\overline{X}} + (A_2 \times \overline{R})$$

where, $\overline{\overline{X}}$ = the process average, or the average of averages

$\overline{R}$ = the average range, or average of the ranges

A_2 is found in the table above

To calculate the control limits for the Range chart, use the following two formulas:

$$\text{Upper Control Limit } (R_{UCL}) = D_4 \times \overline{R}$$

$$\text{Lower Control Limit } (R_{LCL}) = D_3 \times \overline{R}$$

where, $\overline{R}$ = the average range

D_4 and D_3 can be found on the table above

Table 2.2 The factors and formulas for calculating control limits

the steel rule and measuring method used during the process potential study. The measurements for overall length would be noted on the control chart paper.

After 25 columns of data were completed, the SPC Coordinator used the information to calculate the control limits. This procedure was performed with a couple of the board cutters present so they could see how the control limits were created. Once the values for the control limits were calculated, one set for averages and one for ranges, a new control chart was drawn. Using the same type of X-bar/R charting paper, the SPC coordinator drew a baseline, upper control limit and lower control limit for average and then for the range.

Step 7: Post the Chart for 30 Days

Once control limits have been established, the chart is posted at the process site. It will take 30 days of production before the stability, predictability, and capability of the process can be determined. However, the chart is not forgotten because it is during these 30 days that most problems with a control chart will occur. These problems are addressed in Chapter 5. The best procedure is to inform the operators that this chart is not "carved in stone." Management should make the workers aware that problems occur frequently and that the first charts will be learning experiences. For noncontinuous processes, the control chart should reflect each start and stop of a production run. Stability, predictability, and capability should be evaluated after five or six charts are completed.

When a plotted point is outside of the control limits, the operator reacts by noting what seemed to be wrong and what actions are necessary to correct the situation. A formal reaction plan is a part of the SPC system and is described in further detail in chapter 4. Part of this reaction to points "out-of-statistical-control" is to isolate the involved production. Usually, the products produced since the last control chart check are put on an "in process hold" condition. This production is then physically isolated from the process. A typical arrangement is to move a tub of parts a few feet to an isolation square painted on the production floor near the process. Any product in this square cannot be moved until a final disposition is arranged. Usually, quality assurance people will examine the isolated material to determine its usability. It is the operator's responsibility to ensure that any questionable material in this isolation group of products is given further examination. The control chart will indicate which parts should be questioned for quality.

The saw operator at 2:00 p.m. on December 3, randomly takes five
shelves and measures them. He obtains the following readings, 0, -2, -4,
6 and 2. This creates an average of 0.4 and a range of 10. These results
are recorded on the control chart in the data area.

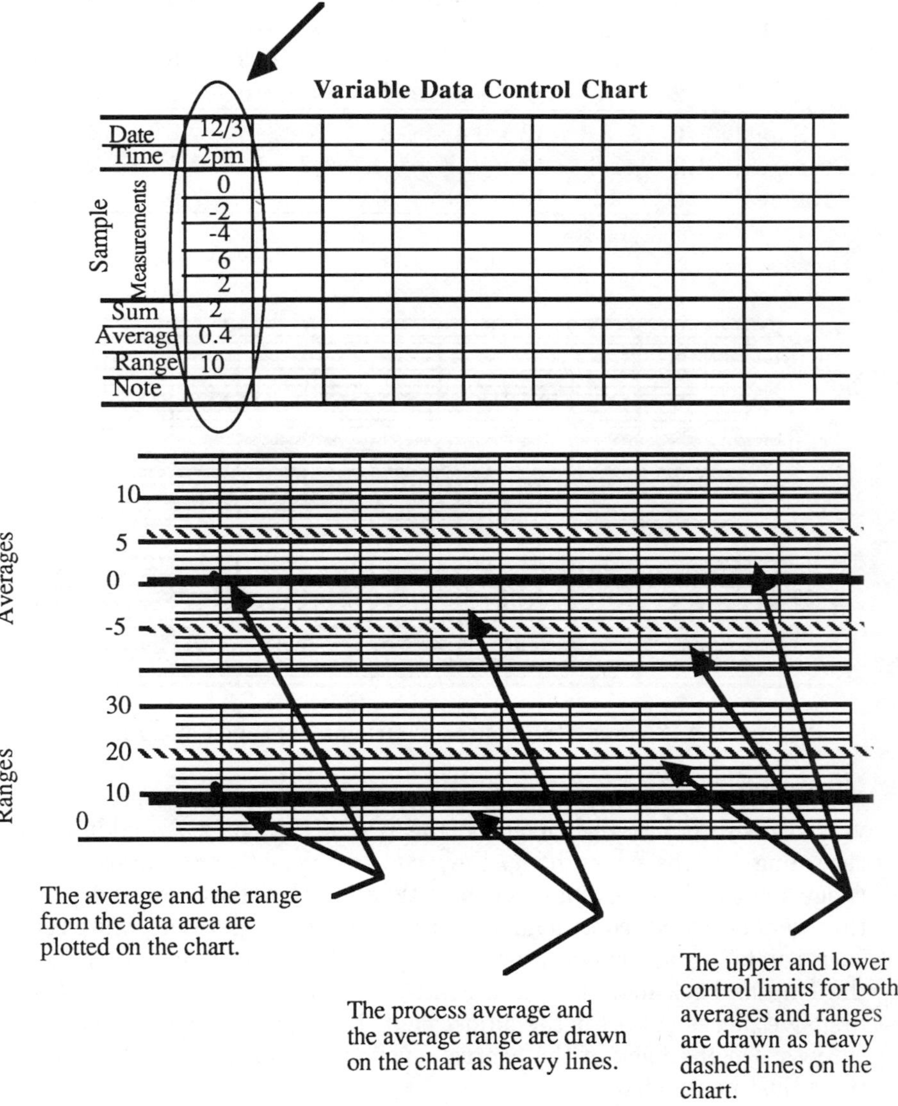

The average and the range
from the data area are
plotted on the chart.

The process average and
the average range are drawn
on the chart as heavy lines.

The upper and lower
control limits for both
averages and ranges
are drawn as heavy
dashed lines on the
chart.

Figure 2.8 Filling in a control chart.

Time	8	9	10	11	12	1	2	3	4	5
1	0	4	2	2	0	2	4	−2	−2	−2
2	4	0	4	0	8	2	2	−2	0	6
3	−4	−6	−2	0	−8	−4	−4	2	−6	0
4	4	0	8	2	2	−2	0	6	−4	0
5	2	−6	0	−8	−4	−4	−6	−2	0	2
Sum	6	−8	12	−4	−2	−6	−4	2	−12	6
Average	1.2	−1.6	2.4	−0.8	−0.4	−1.2	−0.8	0.4	−2.4	1.2
Range	8	12	10	10	16	6	10	8	6	8

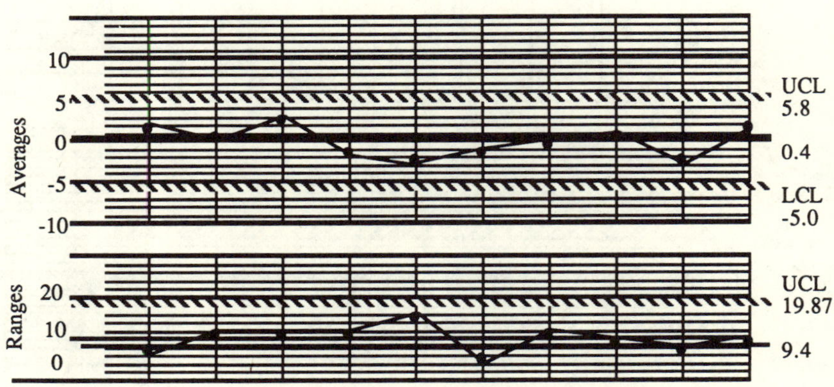

Figure 2.9 A completed control chart for averages and ranges.

The Xeron company posted its new X-bar/R chart only after all involved operators were briefed on the use of the chart. In addition to recording lengths every hour, they had to calculate an average and a range for the five readings and plot them on the chart. Plotted points that went outside of control limits were noted by the operators who circled such points. Then the operator was expected to adjust the saw to bring the lengths back into statistical control. By checking the length of the first five pieces cut after the adjustment, the operator could confirm its success. On the back of the control chart, the operator would then list the adjustments made.

As part of this new SPC system, the cutting area supervisor also started to maintain a process log for the area. In this log, records were kept of when saw blades were changed, new wood was received, or any

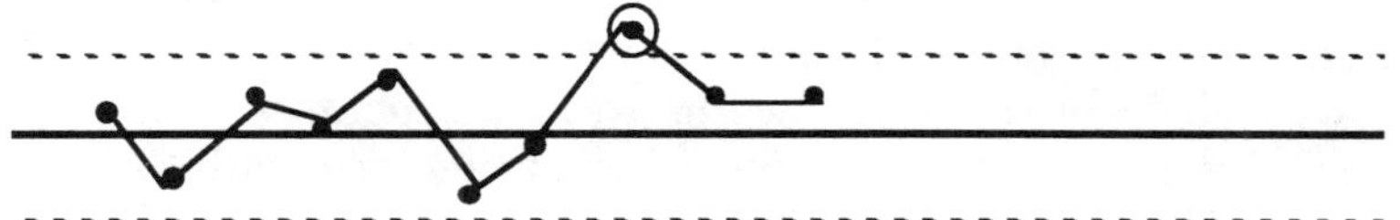

Points on a chart that are out of statistical control should be circled.
The chart above has lost statistical control by crossing a control limit.
The chart below has a identifiable trend occuring.

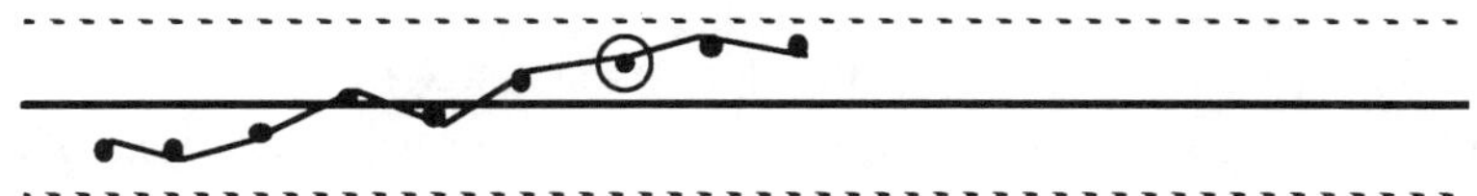

Anytime a point on a chart goes out of statisitical control, the operator
should flip the chart over to the control chart log on the back. Here the
date, time, problem, actions taken and the solution should be noted. A
supervisor should sign all entries.

Control Chart Log				
Date	Time	Problem	Actions Taken and Solution	Init.
12 / 3	2:00pm	Lengths became too long	Guide tightened (no change)	
			Saw Arm tightened (worked)	RC
12 / 5	12 am	Upward trend on the chart	Fixture found to be loosening	SB

Figure 2.10 Loss of statistical control and how to note its correction on
the back of the control chart.

other changes were made in the process. This information would be used
later to help locate potential causes of problems with the process. For
example, blade wear may show up on the control chart as a steady trend
in the averages; an increasing trend in the range may indicate a loosening
of the guide.

Meaning of Out-of-Statistical-Control. Another phrase that is commonly used instead of "out-of-statistical-control," is, "out-of-control". An example will illustrate the difference between the phrases. One day a driver was attempting to lift a ten ton coil of spring steel with the forks of a lift truck. The bands of the coil caught on the forks and broke. The spring steel began to unwind in a spectacular fashion. In fact, it began to unwind right through a brick wall and several nearby cars. The lift truck operator identified the situation correctly with the phrase, "out-of-control" and immediately implemented a reaction plan. The driver leaped off the lift truck, ran in the opposite direction from the steel, cleared people from the area, and began thinking about what to tell the boss.

"Out-of-statistical-control" refers to a less serious situation. The board cutters from the previous example have been consistently getting five board averages within established control limits. At 4:10 p.m. on Thursday, they take five randomly selected boards and the average is outside of the control limits. They use the correct phrase "out-of-statistical-control" and implement their own reaction plan. The phrase is to indicate that the process seems to have changed and that investigation is necessary. In this case, the average length is too long. Past experience has taught them this is usually caused by the guide failing to hold the boards tight, resulting in cutting boards too long. A quick check confirms this is so. The guide is tightened and production returns to normal conditions. When corrections have been made, the process is called "in-statistical-control". A control chart indicates that a process is "in-statistical-control" when all observed averages and ranges, proportions, or average counts fall within the control limits.

A Finer Interpretation of Control Charts: Zone Analysis. A process can remain within the control limits and still have lost statistical control. Zone analysis is a method of detecting subtle trends or patterns on the quality control charts. To perform one zone analysis, the area between the control limits on the control chart is divided into three equal zones. The center third of the area is one zone. The easiest rule to remember with this center zone is that two-thirds of all points plotted should be within this zone.

The illustration above shows the zones and details the rules to remember. One that is not included is a run of seven points in the center zone. One possibility is seven points in a row plotted on the process average. This is a suspicious pattern. If someone were cutting a deck of cards with a friend for a dollar a cut, and cut seven aces in a row, would this be luck, or is something peculiar going on? In this situation, confirm that the operator is plotting accurately and check the gage in use to see

There are situations where points on a control chart will not cross the control limits, but statistical control is still lost. Unusual patterns on control charts indicate a loss of statistical control. What zone analysis does is provide some easily remembered rules for spotting unusual patterns. First, divide the area between control limits into three zones.

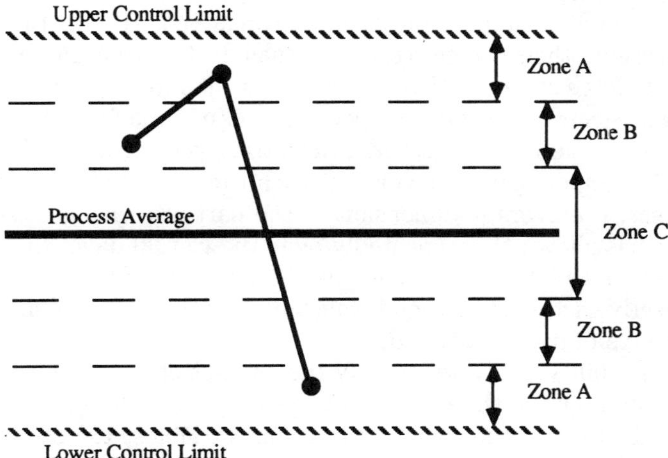

The easiest rule to remember is the one-third, two-third rule. It states that two-thirds of all the points on a control chart should be within the center one-third of the control area. In this case, within Zone C. Such a condition reflects a process in statistical control.

The second rule is the Zone A rule. The outer third of the control area is defined by the two Zone A's above. Statistical control has been lost whenever

2 out of 3
or,   3 out of 7
or,   4 out of 10

points are in Zone A. The 2 out of 3 condition occurs most frequently. The points on the above chart illustrate one possible condition of "2 out of 3".

Also watch for 4 out of 5 points in Zone B or 8 points in a row in Zone C.

Figure 2.11 Zone analysis, a method for detecting loss of statistical control.

if it somehow is stuck or not capable of measuring the amount of piece-to-piece variation involved.

Step 8: Second Capability Study

After 30 days of statistical control (that is, all points staying within control limits) a second capability study can be conducted. If statistical control is not present, then corrective actions should focus on gaining statistical control. A chart in statistical control is monitoring random causes of variation. Sometimes there are special causes of variation. These are the ones that can readily be identified as upsetting your process. For example, a metal stamper may discover that each time a roll of steel is changed on a press, the average dimension on the part also changes. If the control chart has shown this repeatedly over the last 30 days, then steps should be taken to try and use more consistent steel. If this is not possible, or corrective actions fail, then recalculation of the control limits based on the 30 days of data is advised.

The second capability study will certify the amount of variation the process produces and will be the benchmark against which all improvements will be evaluated. In the case of charts for variables, the study involves the calculation of the Cp and Cpk indexes. For attribute charts it is expressed as the baseline.

The Xeron company found that reaching statistical control was a difficult road. The first thing that went wrong with the new X-bar/R chart was that the averages were soon fluctuating wildly across the chart. Several days of studying the problem revealed that sawdust accumulating in the guide would throw off the length of cuts. The operator would react by adjusting the guide, which only increased the variations in averages. The problem was finally solved by installing a small broom next to the saw so the operators could periodically remove the sawdust from the guide.

Regular removal of sawdust made the plots of averages more stable. However, "out-of-statistical-control" points remained frequent. Only after several weeks of converting to a system of regular blade changes, tightening of the guide, and purchasing wood of consistent quality did the process become stable and predictable. Finally, the Xeron group was ready to calculate capabilities within the process while it was under statistical control.

By calculating the Cpk ratio of 0.725, the Xeron team was able to determine that the process was stable, predictable, and capable. They had achieved statistical control of the process. The consistency of the cut board could be guaranteed. Upon examining the scrap reports for the bookcase line, the team discovered that the scrap rate had fallen by two percentage points. The economic success of the control chart could be documented to management. In addition, the operators of the saws had learned a great deal about the conditions that make their process successful which increased their skills and worth to the company.

Step 9: Make Corrections/ Updates

It is very rare that the first control chart will find the process is on target or exceeds specifications. An example is the presence of tool wear. If charts show the slow and steady wear on tools, then perhaps it would be possible to change the baseline to a position that would take advantage of the wear, that is, reset the process so that the baseline is slightly lower than the optimal specification. Then, as the tool wears and the parts become larger, there is more time between tool changes.

For the Xeron team, the success of its control chart is not the ending, but a signal of the beginning of its involvement in the SPC system. The chart is used forever. As new changes in the process are introduced or alterations in methods are suggested, they can be evaluated directly by examining their effect on the control chart. For instance, one day the averages suddenly became extremely steady. The inner zone of the control limits had all of the plotted points. A quick check by the supervisor revealed that the steel ruler used for measuring had been lost the night before and in its place, an inferior wooden ruler was being used. The new ruler was only marked to the sixteenth of an inch. Variations were not being effectively measured; the control chart had detected this variation and a new steel ruler was quickly purchased.

Step 10: Use of the Chart for the Continuous Improvement of the Process

Continuous improvement of a process is the central philosophy of SPC. Continuous improvement can also be a radical departure from the conventional wisdom about how industrial processes should operate. Tradition held that the goal of any process was to meet customer requirements. The philosophy of continuous improvement holds that profitable advantages accrue in developing a process beyond customer requirements. There is a competitive advantage to a process that exceeds customer requirements resulting in more consistency among the products and more margin for errors. In addition, once a process is stable, predictable, and capable, it is very advantageous to the company to explore ways to make a faster, cheaper process. In short, statistical control is needed for continuous improvement. Without a baseline and an expected range of variation, it is difficult to evaluate real progress.

The Cpk Index checks both the effect of process variation and how well-centered the process average is.

First, draw a picture of the tolerance spread and where the process variation is in relation to the specifications.

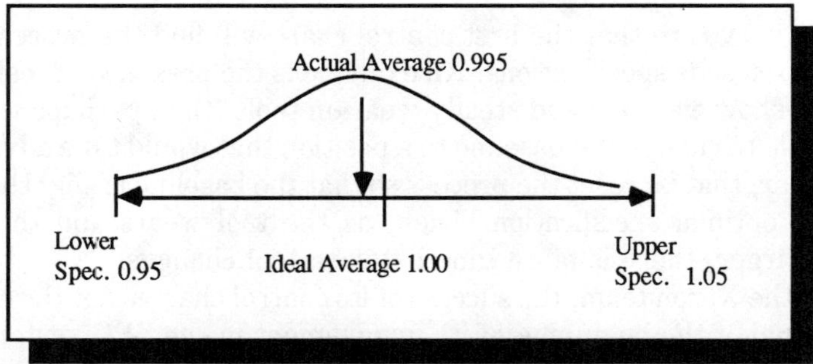

Calculate the Cpk Index by choosing which spec. limit the actual average is closer to.

$$\frac{\text{Upper Spec. - Actual Average}}{3 \times \text{Standard Deviation}} \quad \text{or} \quad \frac{\text{Actual Average - Lower Spec.}}{3 \times \text{Standard Deviation}}$$

The ideal process has a Cpk of 1.33 or higher.

The data from Table 2 - 1 indicates an average of zero and lower specification of negative eight. The standard deviation equals 3.67658. Therefore, the Cpk index would be calculated as follows:

$$\text{Cpk} = \frac{\bar{\bar{X}} - \text{LS}}{3 \times \text{s.d.}} = \frac{0 - (-8)}{3 \times 3.67658} = \frac{8}{11.02974} = 0.725$$

A Cpk index of 0.725 is below the minimum requirement of 1.33.

Figure 2.12 Formula and calculation of a Cpk index of capability.

The last step is the most difficult of all: using the control chart for the life of the job. The chart must become a regular part of the process. It is the responsibility of the operator to keep and maintain the chart in a timely manner, while management should use control chart data for making decisions about ways to improve the existing process. Unfortunately, once the chart is posted, many SPC people tend not to check on its progress. The line supervisor must check the charts daily for completeness and accuracy. If the line supervisors do not support the SPC system, then the expected life span of a control chart is very short. All too often, control charts on a shop floor are one or two weeks behind production. They must be current or they are useless.

The Xeron team met to discuss the transfer of responsibilities to the cutting room operators, and decided to form the operators into a problem-solving team. This new cutting room team was trained in problem-solving methods. The SPC Coordinator was part of the team and provided direct support, while indirect support came from the original management team. The cutting room team met with the management team every quarter to discuss what improvements to try.

One of the first proposals was to meet with the engineers and the purchasing people to discuss the type of wood that was being bought for the bookcase. The cutting room team convinced the engineers and purchasers that a warp resistant wood would reduce the number of customer complaints. The purchasing people changed their policies about wood buying. In the future, all contracts would only go to those suppliers who had demonstrated the warp resistance of their materials. The cutting team later noticed the effect of this change on their control chart. The capability of the process increased significantly. New control limits were calculated and the purchasing group was congratulated.

NO FINAL STEP: PREVENTION VERSUS DETECTION

In a complete SPC system, there is no final step in the use of a control chart. If the control chart's use was to end with only the real-time monitoring of a process, then the SPC system is designed for *detection* of problems. A true SPC system goes beyond this and strives to *prevent* problems. This prevention method is created when the chart, its related capability studies, and the process log are used together. The process log is particularly important since it records all adjustments and changes the process experienced.

Past problems are studied in detail. The main question is what it would take to have prevented the problem in the first place. A work team discovered that a tool loosens regularly on a machine, thus causing increases in part-to-part variations. By either adopting a regular schedule of tightening the tool or designing a new tool holder, the problem can be prevented. Preventing a problem is much more economical to a company than detecting a problem and correcting it.

PAYING TOO MUCH ATTENTION TO CONTROL CHARTS

The chief criticism of control charts is that they force operators to pay too much attention to process stability and not enough to realities of the environment. In the case of the Xeron Furniture Company, it would be easy to imagine a situation where operators sacrifice productivity for problem-solving meetings. Perhaps this team is not maintaining a good rate of production while trying to obtain statistical control. Many companies point out that some fast processes cannot be interrupted, therefore operators cannot stop to fill in control charts.

For these situations, alternatives based on the principles of SPC are used. A basic principle of SPC is that the operator must have direct access to the monitoring of a process. If the process is one that the operator cannot chart, then possibly a line supervisor or inspector can do the actual charting. However, the chart must stay near the operators so they can be kept apprised about the status of the process.

Another related issue is that the SPC system tends to focus efforts on a single dimension or characteristic. Other dimensions have to have some form of control. A good example is a plastic extrusion for a window molding. A particular dimension may be critical for the operation of the window, but adjusting the machine to perfect that one dimension will alter the other dimensions of the extrusion.

The answer to these concerns is in extensive planning and flexibility in procedures. SPC is a new technique for almost all companies and changes are inevitable. With change there will be resistance and some harmful effects to the processes that must be anticipated and avoided where possible. When not possible, the workforce must be made aware of the dangers. Operators of the extrusion example above could be told to maintain one dimension as best as possible, but to also monitor the other dimensions. At the same time, a quality assurance inspector or engineer could be on hand to assist in the stability of the process while the control chart is under development.

SUMMARY

The control chart is a competitive tool only when it is used wisely and honestly. The company that wishes to implement control charts must do so in a timely manner after trusted lines of communications are established between management and operators. It is management's job to effect the implementation of control charts. Quality assurance is a facilitator of the process and should never be given sole responsibility for the control charts. SPC belongs within the production control function (see Chapter 8). In the end, the operators should have the responsibility of filling in and using the charts; quality assurance, the responsibility of setting up and monitoring charts; and management, the job of planning and auditing the system.

The control chart is a simple statistical mechanism for detecting deviations in an existing process. Significant deviations are seen as either points outside of the control limits or as unusual patterns on the chart. However, without a human reaction to these points, a control chart is a worthless scrap of paper. The information from a control chart must be acted upon and used to improve the operations of a process and its related procedures. In a complete SPC system, the information from control charts is used for expanding the competitive position of a company.

CHAPTER 3
ATTRIBUTE CONTROL CHARTS

Many times the critical characteristic on a product cannot be measured by a variable scale. Perhaps the characteristic is a visual defect or other form of flaw. In these cases, an attribute control chart is employed. An attribute chart monitors the average number of defects or defectives within a group of products. These charts are not used the same way as a variables control chart; statistical control is a momentary goal. The real purpose of these charts is to supply the information needed for eliminating the causes of defects or defectives.

DEFECTS VERSUS DEFECTIVES

A defect, also called a nonconformance, is a flaw of some sort in the product. A defective, or nonconforming product, is one that does not meet quality standards. In other words, it is judged unfit for its intended use. Defects come in many shapes and forms. The Xeron company case from the previous chapter can further illustrate these points. In the Xeron assembly process, there was a finishing area where lacquer was applied to the assembled and stained bookcases. How many defects could occur to just the lacquer finish? A list would be almost endless, including defects such as: cracking, runs, drips, crazing, scratches, bubbles, fogging, and so on. Sometimes the defects would either be so severe or numerous, the entire bookcase was judged to be defective.

From the customer's point of view defects can have a range of importance. For example, military standards generally divide this range into three parts.

1. Critical—the defect is harmful to either those making the product or those using the product.
2. Major—the defect will affect the ability of the company in selling the product.

3. Minor—the defect is annoying, but will not affect the saleability of the product.

Most companies go to great lengths to avoid critical defects. These usually result in lawsuits, damaged reputations, and the catastrophic economic effect on the company. Major and minor defects used to be the concern of quality assurance. Acceptance by lot sampling was the common method for making sure that too many defects did not get shipped. Defectives in the form of scrap, rework, and customer returns were the concern of management and production. Under an SPC system, all of these situations are the concern of all employees. The control charts form the front line in the battle against defects and defectives. As Chapter 10 will discuss, the final battle for quality is waged within the design.

The reason for concern about defects and defectives comes from the economics of the times. Scrap is the best example of this impact. When a product is defective it must be scrapped or reworked. In the past, most companies accepted scrap as part of the cost of manufacturing. It was not unusual for a plastics engineer to add 10% to a bid for a part to account for the expected scrap. In the late 1970s many companies began to evaluate the cost of scrap as an unnecessary expense. Many found that scrapping a part is over seven times more expensive than producing it right the first time. The costs came from having to pay someone to move, sort, and dispose of the parts, the price of wasted material, and the loss to production from having to make the part again. Therefore, today's emphasis is on zero scrap and zero defects. If a product is made right the first time then, the cost of production is lowered, and a company can compete more efficiently. As will be discussed in later chapters, zero defects are needed for implementing other manufacturing technologies.

ESTABLISHING AN ATTRIBUTE CHART

The steps toward implementing an attribute chart are the same as those for a variables chart (see previous chapter). Naturally, the calculation of control limits differs according to the type of chart being used. The capability studies are still performed, but larger samples are taken. Attribute data contains less "information" than variable data, so more pieces have to be examined. Below, the Xeron example will again be used to illustrate the implementation of an attribute chart.

There are four types of attribute charts. They are called the p-bar, np-bar, C-bar, and u-bar charts. Two questions are asked to determine

Selecting the Correct Attribute Chart

Characteristics being examined

		Defectives	Defects
	Consistant	np - chart	c - chart
Sample Size	Variable	p - chart	u - chart

where,   p = the percent defective, calculated as - $\dfrac{\text{parts found defective}}{\text{sample size}}$

np = the number of defectives

u = count of defects per unit

c = count of all defects in a sample

The proper attribute chart is selected by asking two questions. First, does the critical characteristic involve defects or is the main concern whether the product is defective? Second, can consistant sample sizes be obtained? If not, then charts for variable sample sizes must be used.

Figure 3.1 How to select the proper attribute chart.

which of the charts will be used. The first question is, "Are defects or defectives the main concern for the process under study?" The second question is, "Can consistent sample sizes be obtained from the process?"

Once chosen, the attribute charts still have some special requirements. The first has to do with sample size. A chart for defectives is dealing with an either/or situation. Either the part is defective or it is not. Either a defect is present or it is not present. This is called attribute information and requires large sample sizes for statistical accuracy. At least 50 units per sample are recommended. When possible 100 or more are desirable.

Luckily, a lot of data involving defectives has already been gathered from large samples of the existing production procedure. For example, scrap reports are based on entire shifts of production. Once established, charts for defects can be used with very small sample sizes. This is especially true with products that have many defects per unit, such as a written report with several spelling and grammar errors.

Another consideration concerns the p-bar and the u-bar chart. With both of these charts the sample sizes can vary. Whenever the sample sizes vary by more than 25%, the control limits must be recalculated for the affected time periods. For example, when the sample size increases, the zone of chance (control limits) decreases.

THE XERON FURNITURE COMPANY

At the Xeron Furniture Company, the original management team has completed its implementation of an X-bar/R chart for the cutting area. Review of the team's notes about the process indicates that defects in the bookcases are the second most common complaint after warping. Since the warp problem seems to be under control, the team elects to attack the defect problem. Further study shows that most of the defects reported are in the finish on the bookcase. Investigations conclude that inspection of the finish takes a good deal of time. Therefore, the team plans for one bookcase to be examined carefully ever hour. A consistent sample size and a chart for defects are specified by the team. From the above decision block, the c-bar chart is appropriate.

At the same time, the team also discusses the internal expenses caused by the scrap seen on their first tour of the entire bookcase line. The team finds that scrap reports are filled out by each section of the line and submitted to the production manager who uses the information in the weekly cost report. The number of bookcases involved changes daily, meaning that sample sizes will vary and that defectives are being monitored. From the above decision block, the team chooses the p-bar chart to monitor overall scrap on the bookcase line.

REVIEW OF THE STEPS TO IMPLEMENTATION

Chapter 2 outlined 10 steps that are taken to implement a control chart. These steps will be reviewed below with the special considerations for attribute charts noted.

Step 1: Meet the Process

The Xeron team has already toured the entire bookcase line to get a feel for the people, environment and machinery involved (see Chapter 2).

Step 2: Select the Critical Characteristic

The first step in selecting a critical characteristic is defining the attributes of interest. The traditional form of this definition is the *defect checklist*. A defect checklist is a written description of all the possible defects for a specific product. Each defect is then ranked. For example, under the military standard system defects would be ranked as either "critical", "major", or "minor". These descriptions are time consuming to create, but later are very valuable to the company. The purpose is to establish the standards for evaluating defects to prevent people from frequently changing the definition of a defect. At one company a quality assurance technician described this phenomenon as the "daily quality standard." To emphasize the point, a roulette wheel was placed in the quality lab marked with areas such as "accepted", "rejected", and "use as is."

The Xeron company team created a defect checklist with the assistance of the finishing people. Then to reinforce the standards, the team hired a photographer to photograph the various defects at their different levels of severity. The color photos were collected into a book that anyone could use as a guide to describe various defects. The team also took a critical step by having the finishing room crew select a bookcase with a perfect finish. This took some time, but finally the crew discovered a bookcase with a finish they described as excellent. This example of the proper bookcase finish was then placed in the quality lab next to the finishing area to reinforce the idea that the company was emphasizing good production over defect detection.

Step 3: Study Specifications and Set-up

The Xeron team watched how the finishing crew worked on a bookcase. The specifications called for a defect free finish with complete coverage by the stain and lacquer. The stain was applied by hand by crew members using cotton rags from a large box kept near the staining area. These rags were used until any member noticed streaking in the stain. The lacquer was applied with a spray gun inside a special spray booth. The gun had to be cleaned whenever it began to spatter the lacquer. Fixing stain problems was relatively easy to correct if they were caught before the lacquer was applied. There were no procedures for fixing a lacquer problem short of marking the bookcase defective.

Step 4: Find the Baseline

With an attribute chart the baseline is the average number of defects or defectives that are detected. In the case of a p-bar chart, this average is expressed as the average percent of bookcases that are defective. On an np-bar chart the baseline is the average number of defective products within a set sample size. The c-bar chart's baseline is the average number of defects found in a given sample size. The u-bar chart expresses this as the number of defects per unit on the average. Examples of these calculation are in Figure 3–2.

The Xeron group began with the c-bar chart proposed for the finishing area. As a process potential study, the team instructed the finishing crew to inspect one bookcase every hour for defects. Naturally, the crew was trained in the use of the defect standards that the company had created. However, to add some fun to the task, they were told to pretend they were the nastiest customers that ever existed and were encouraged to make a contest out of how many defects they could find on a bookcase. This would ensure that the team would have standards higher than any customer. After all, the crew would know exactly where and how to look for defects.

On a piece of paper they recorded the information that would form the process potential study. On one side of the page was a list of possible defects. After examining a bookcase, the team would tally the number of each type of defect found. On the other side of the page, the team had a drawing of the bookcase from two different views. On this they created a defect map. The most frequent defects detected on each bookcase were marked on the drawings. The idea was to plot the position and type of defects. Over time defects tend to cluster on such maps. These clusters provide clues as to the cause of the defect. For example, the finishing crew discovered stain was missing near the bottom of the bookcase. This was easy to understand, since the stainers had to bend over to cover these areas, and frequently a shadow was cast from the overhead lights.

The total number of defects is recorded at the bottom of each column. To create a c-bar chart, these numbers are then averaged. In this case the crew found that the average bookcase they examined had 6.3 defects. This number is the baseline for the process. What the number means is that on a normal day, a typical bookcase will have about six defects.

The management group has also been busy during this time and has arranged to receive copies of the scrap reports once a week. Each day of production is seen as a sample of the process. Dividing the number of defective products scrapped each day by the total number of products produced, the team calculates "p", the percent of scrap. On each of these

Formulas for calculating control limits:

NP Chart: UCL/LCL = $\overline{np} \pm 3 \times \sqrt{\overline{np}\left(1 - \dfrac{\overline{np}}{n}\right)}$

P Chart: UCL/LCL = $\overline{p} \pm 3 \times \sqrt{\dfrac{\overline{p}(1-\overline{p})}{\overline{n}}}$

C Chart: UCL/LCL = $\overline{c} \pm 3 \times \sqrt{\overline{c}}$

U Chart: UCL/LCL = $\overline{u} \pm 3 \times \sqrt{\dfrac{\overline{u}}{\overline{n}}}$

n = sample size
p = percent defective
np = number defective
c = number of defects

$\overline{p}$ = average percent defective
$\overline{np}$ = average number of defectives
$\overline{c}$ = average number of defects in
      sample
$\overline{u}$ = average number of defects per
      part

k = number of sampling
      intervals

Figure 3.2 Formulas for the attribute control charts.

Defect Map of Bookcase

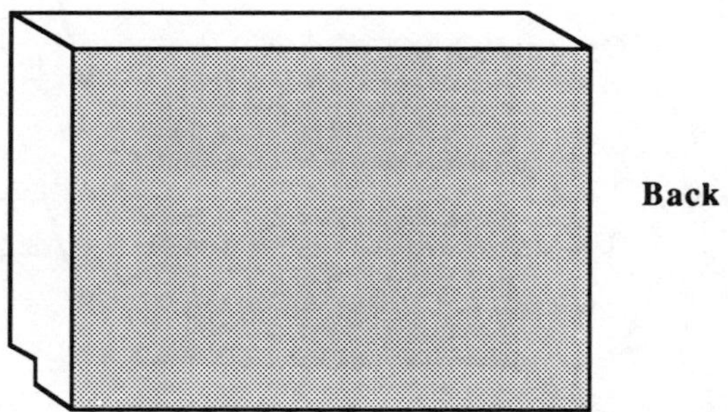

Defect   Checklist

Critical Defects - must be reported immediately

1. Warp - noticable distortion of the shelves.

Major Defects - record on control charts

1. Cracking - visable cracks in the finish or wood
2. Missing Finish
3. Missing Lacquer

Minor Defects - record on control charts

1. Runs - runs in finish or lacquer

Figure 3.3 The defect map and checklist for the bookcase.

Potential Study for Defectives

Company : Xeron
Division : Bookcase Assembly Line
Date : December 8

Number Sampled	Number Found Defective	Percent Defective
110	12	11
100	18	18
100	22	22
100	18	18
120	22	18
110	21	19
100	23	23
100	15	15
100	24	24
110	13	12
100	14	14
150	18	12
100	23	23
100	30	30
106	17	16
100	15	15
100	14	14
100	17	17
100	20	20
90	19	21
100	15	15
100	15	15
100	12	12
100	22	22
100	17	17
2,596	**456** Total	

$$\frac{456}{2,596} = .1756 \text{ or } 17.56\% \text{ average scrap rate } (\bar{p})$$

Average sample size is 104

Figure 3.4 The process potential study for a p-chart.

days the team is careful to note what made the bookcase defective. This information will later help in the finding and eliminating of the causes of scrap. Reworked items are also listed as part of this scrap chart.

The percent of scrap each day is recoreded for 25 days. These 25 percentages are then averaged to form the p-bar, or average percent defective. In this case the result is 0.1756. This is the baseline of the scrap rate at Xeron's bookcase line. In other words, on a typical day, the bookcase line will have to scrap 17.56% of production.

Step 5: Estimate Expected Variation for Creating Control Limits

The variations in the p's, c's, u's, and np's for each chart are expressed by their control limits. The c-bar chart is one way to illustrate this. The number of defects detected each time will vary and the amount of variation due to chance is expressed by the control limits. It is pointless to worry about the variation in the number of defects for individual bookcases within a lot. The existence of any defects is important, but it would be too costly to sort after careful examination of each bookcase.

Step 6: Create the Control Limits.

The Xeron teams calculated the control limits for both the c-bar and the p-bar charts.

The c-bar control limits mean that whenever one bookcase is examined, the minimum number of defects expected is zero, and the maximum number expected could be as high as 13.83. If numbers different from these are found, then a change has taken place in the process. Higher levels of defects require immediate action to correct the process. Lower levels of numbers mean that something has improved the process. The source of improvement must be found immediately. Therefore, any points outside of control limits are treated in the same way as an X-bar/R chart. Again, the problem and its solution are noted.

The p-bar has control limits at 29% and 6%, indicating the range of normal scrap rates for the process. Rates higher or lower than these limits indicate that the process has changed. Again, higher rates require correction and lower rates require investigation for improvement. With any attribute chart, the rules of zone analysis also apply. Trends and unusual patterns call for a documented reaction.

Both of the above charts list possible defects. After each sampling, a tally should be kept of the defects that were detected or caused a product to be judged defective. This information will be invaluable later on.

The work team begins the creation of a c-bar chart by carefully examining
a group of bookcases and recording the number of defects found.

Defects	Tally	Total
Warp	//////	6
Cracking	/////////////////	17
No finish	//////////	10
No lacquer	////////////////	16
Runs	///	3
Drips	////	4
Crazing	/	1
Scratches	/	1
Fogging	/	1
Other	////	4
Total Pieces Examined : 10	Total Defects	63

By dividing the total number of defects found by the number of bookcases examined,
the work team finds a c-bar of 6.3 (the average count of defects per unit). The c-bar
is then used to calculate the control limits.

$$\text{UCL / LCL} = \bar{c} \pm 3 \times \sqrt{\bar{c}}$$

$$= 6.3 \pm 3 \times \sqrt{6.3}$$

$$= 6.3 \pm 7.53$$

thus, The UCL equals 13.83 and the LCL equals -1.23,
since this is impossible, the LCL is seen as zero.

Figure 3.5 The process potential study and control limit calculations for
a c-bar chart.

Step 7: Post the Chart for 30 Days

As with the X-bar/R chart, the attribute charts are posted in the
appropriate work areas for at least 30 days to see if the control limits
are good predictors of process capability. During this time, any obvious
problems should be addressed and corrected. Again, the objective is to

Using the information from Fig. 3 - 4, the control limits for the management work team's p-chart can be calculated.

$$\bar{p} = .1756 \qquad\qquad \bar{n} = 104$$

$$\text{UCL} / \text{LCL} = \bar{p} \pm 3 \ \times \ \sqrt{\frac{\bar{p}\,(1 - \bar{p})}{\bar{n}}}$$

$$= .1756 \pm 3 \ \times \sqrt{\frac{.1756 \times .8244}{104}}$$

$$= .1756 \pm 3 \ \times \ 0.0373$$

$$= .1756 \pm .1119$$

Upper Control Limit = .2875 (about 29%)
Lower Control Limit = .0637 (about 6%)

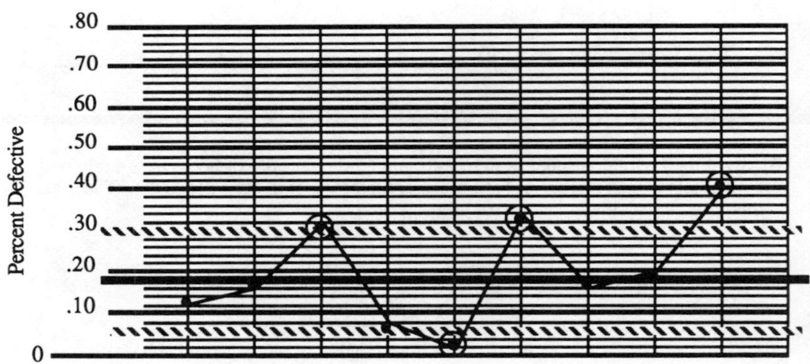

No. Checked	90	100	100	110	95	100	100	100	100
No. Defective	11	18	30	8	4	32	18	19	40
% Defective	.12	.18	.30	.07	.04	.32	.18	.19	.40

Figure 3.5b The calculation of control limits for a p-chart. Note how the chart in use has no statistical control.

Using the information from Fig. 3 - 5, the bookcase work team can create a c-chart and begin filling it in.

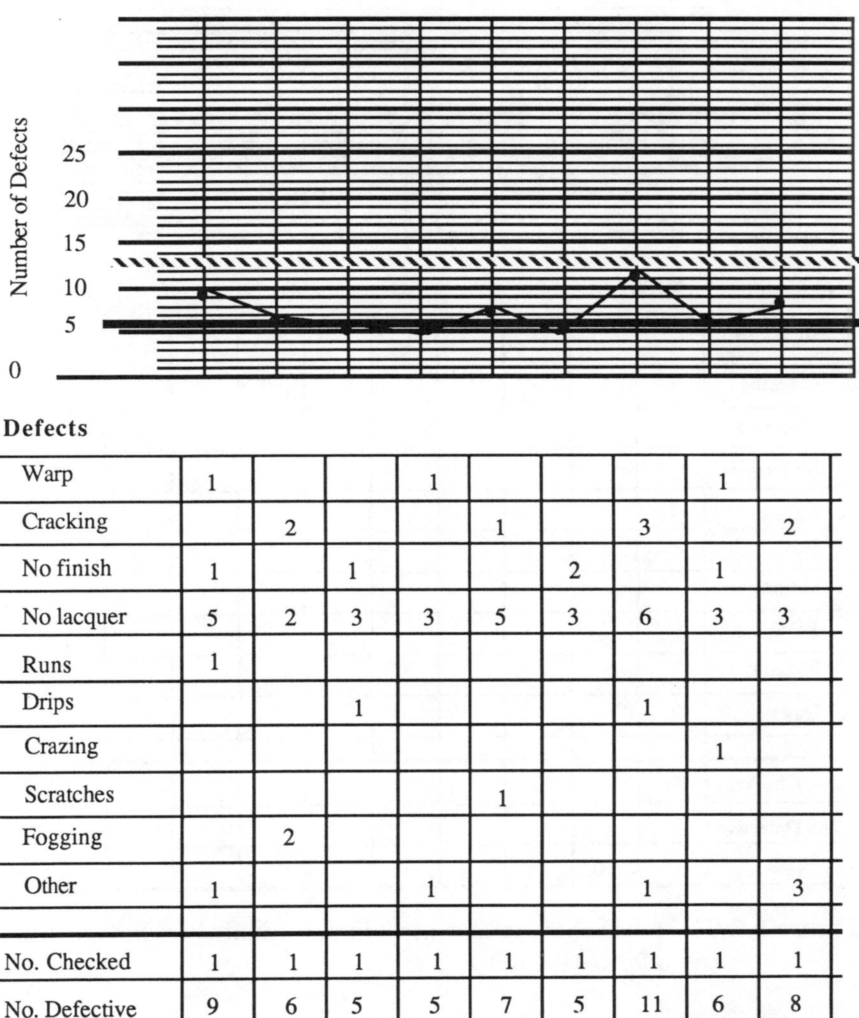

Defects

Defects									
Warp	1			1				1	
Cracking		2			1		3		2
No finish	1		1			2		1	
No lacquer	5	2	3	3	5	3	6	3	3
Runs	1								
Drips			1				1		
Crazing								1	
Scratches					1				
Fogging		2							
Other	1			1			1		3
No. Checked	1	1	1	1	1	1	1	1	1
No. Defective	9	6	5	5	7	5	11	6	8
% Defective									

Figure 3.6 A completed c-bar chart in use.

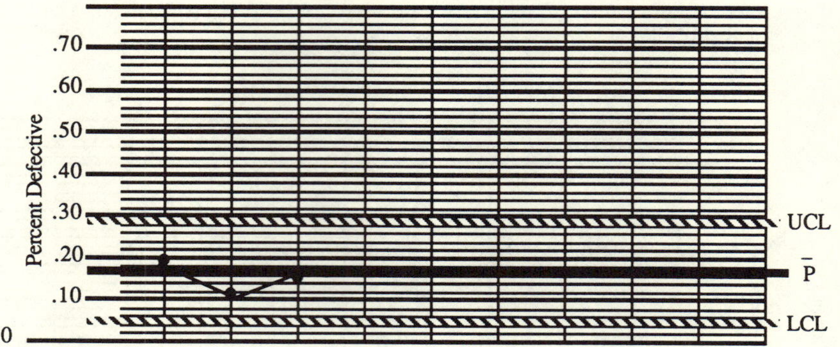

Defects

Warp	1	1									
Cracking	1										
No finish		2	1								
No lacquer	3		1								
Runs	2		5								
Drips		4									
Crazing											
Scratches											
Fogging			1								
Other	2										
No. Checked	50	70	50								
No. Defective	9	7	8								
% Defective	.18	.10	.16								

Figure 3.7 The p-bar chart with control limits. Limits are derived from the information in fig. 3-4.

bring the process under statistical control, which may take from 30 days to one year. Attributes can be difficult to work with and require large amounts of time. As will be shown later, maintaining statistical control will not be the final objective of attribute charts.

When the Xeron teams decided to post their control charts, the results were mixed. The scrap rate fluctuated wildly over the first 30 days. However, the c-bar chart behaved with stability. The finishing crew was

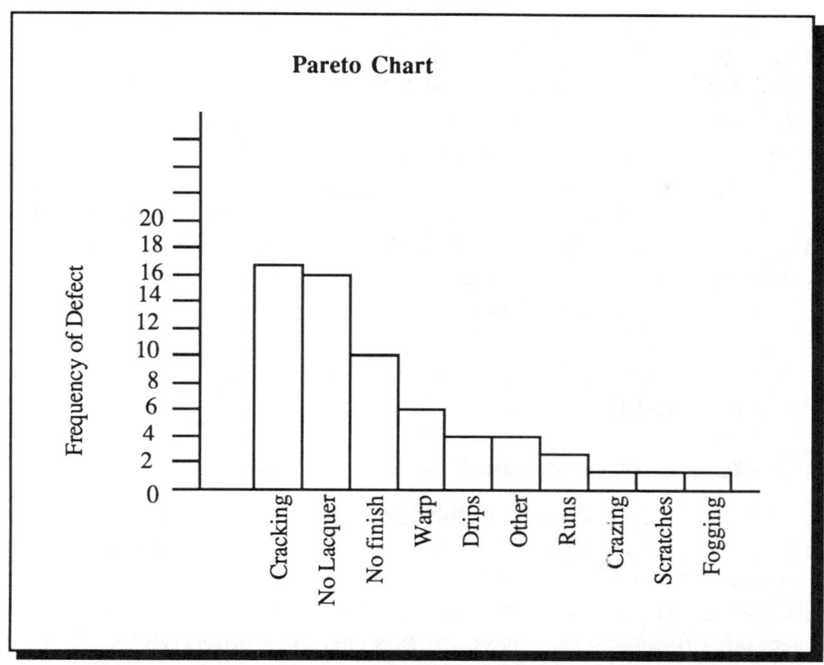

The bookcase work team creates a Pareto chart of defects using the information from Fig. 3 - 5. After improvements, they will create another Pareto chart to compare with this one. See Fig. 3 - 10.

Figure 3.8 The Pareto chart. Defects are sorted by their frequency of occurance.

instructed to proceed with the implementation of its c-bar chart. The management team decided to address the causes of the changes in the scrap rates. Further investigation revealed that the scrap rate would jump whenever materials of marginal quality were used.

Step 8: Second Capability Study

For attribute charts this second capability study only takes place after the process is under statistical control. This capability study is intended to describe the abilities of the process when first brought under statistical control. Then the information on the defects detected are placed in a Pareto chart which is a bar chart that shows the relative frequency of each defect. Taken as a whole, this capability study is the benchmark for evaluation of future problem-solving efforts.

The Xeron finishing crew was able to create its capability study after the 30-day posting period. The management team would have to find and remove special causes of high scrap rates for six months before its p-bar chart demonstrated statistical control. A typical case involved the lacquer spray gun. When it began to malfunction, a bookcase's finish would be ruined. After each adjustment, the spray gun would work for a while and then would ruin another bookcase. Usually, an entire day would go by before the spray gun would begin to work correctly. It was discovered that when the lacquer spray gun was cleaned twice every shift, then malfunctions all but vanished. However, this removed only one cause of the erratic scrap rate.

Step 9: Make Corrections and Updates

Using the information from the second capability study, new control charts with revised control limits were posted. However, there is a major difference in the use of attribute charts versus variables charts. Once statistical control is demonstrated, all efforts turn towards driving the process out-of-statistical-control. In this case, the objective was to force the process into a downward trend. As stated at the beginning of this chapter, zero defects is the standard for modern manufacturing or service industries.

At this point, problem-solving teams are formed. Their job is to begin experimenting with possible improvements to the process. Each successful effort is defined as the chart loses statistical control towards zero defects. This may be seen as a trend of plots heading towards zero, a run of seven points below the baseline, or the plots landing outside of the lower control limit. Once this happens, the new procedure, tool or method is adopted, and the chart is modified. Then new control limits are drawn to reflect the lower level of defects or defectives.

Step 10: Use the Chart for Continuous Improvement of the Process

This is the step where an attribute chart really shines. A reexamination of the control limits will show why. If a team tries a new procedure, the control limits will supply immediate feedback on its effect. If the process plunges below the lower control limit, then the process has been significantly improved. If the process exceeds the upper control limit, the procedure is shown to be harmful. Such information can be valuable to improve a process. Finally, if the process remains in statistical control, then the procedure has no effect on the quality of production. This is

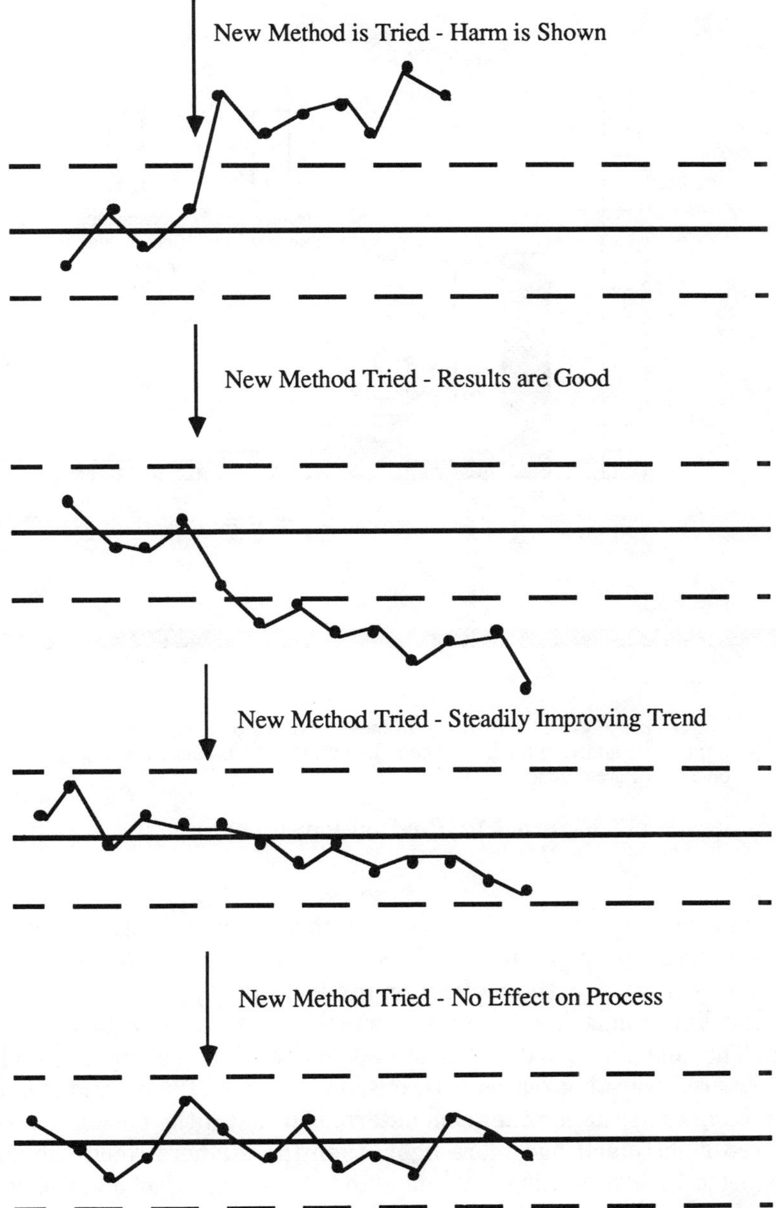

Figure 3.9 Measuring the effect of a new method on an attribute chart.

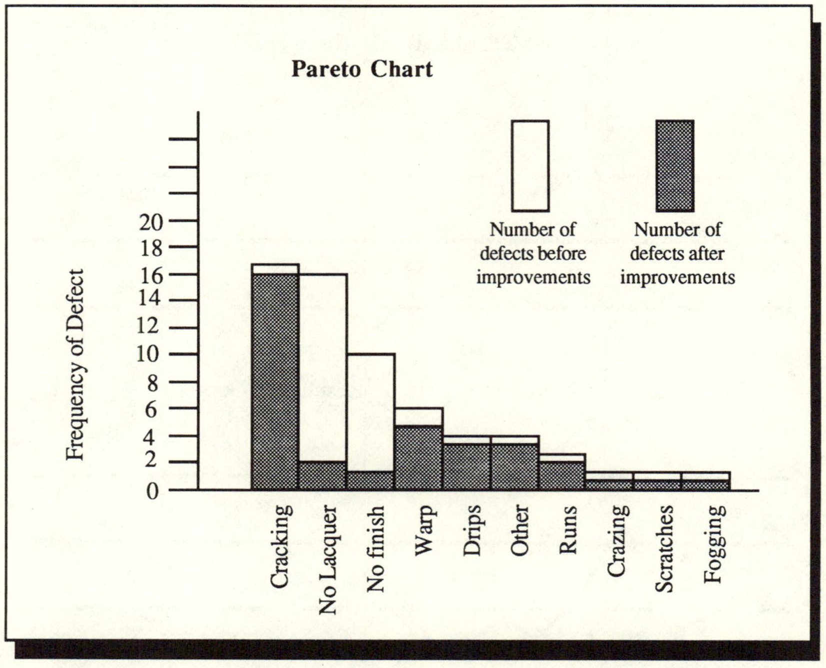

The bookcase work team now examines ten more bookcases after the new lights were added to the line. The reduction in defects found are reported on a new Pareto chart.

Figure 3.10 Revised Pareto chart.

important for other teams since they can then avoid a similar procedure. As an alternative, if the tested procedure costs less to use, it could be adopted as cheaper with no effect on product quality.

The Xeron finishing team was trained to work as a problem-solving team. The defect information had already shown that the largest problem in their area were bookcases with missing patches of stain and lacquer. Examining the situation the team determined that if the bookcases were elevated sightly and had more light, then the stainers could see their work better. Floor lighting was added to the staining room and the bookcases were placed on taller racks built by the team out of scrap wood from the cutting area. The cost of these changes amounted to $2,500. However, the average number of defects dropped on the c-bar chart. The team then estimated that these changes would prevent the scrapping or reworking of 16 bookcases a week. The return on investment would be seven weeks.

The finishing crew then adjusted the chart's control limits to match the new lower levels of defects. Another Pareto chart was drawn to demonstrate the reduction in the total number of defects from correcting the staining problem.

However, things were not going as easily for the scrap rate team. The team soon discovered that the number of products produced daily would frequently change in size, so that control limits had to be recalculated with the changes. Furthermore, attempts to control the scrap rate were proceeding slowly and with great difficulty. Upon calling their SPC consultant, the company learned that this is an indication that the process it is monitoring may be too complex to be controlled by a single chart. The consultant recommended that the bookcase line adopt control charts at each stage of production and that the bookcase line scrap rate chart be used as a barometer of process quality. All related work teams would be told to examine the effect of their area efforts on the scrap rate chart as part of their management report. The management team would then record on the scrap rate chart when new procedures were introduced. The accumulated effect of several problem-solving groups should have an impact on overall scrap.

As each solution is discovered by the work teams, a meeting must take place to determine the best method of continuing the improvement of the process. Two considerations are critical to this process. First, a solution must become a regular part of the process. Some consultants make a living by working for companies that fail to maintain previous solutions. An unsupported solution will eventually fall into disuse. SPC itself is a good example. It was widely practiced in the United States during World War II, but then no longer supported by Department of Defense demands, it fell out of favor in the following years.

The second step is to remove inhibitors and exploit opportunities. The finishing team discovered that some of the stainers did not like the taller racks, so, the area's supervisor will have to make accommodations to encourage their use. On the other hand, the stainers pointed out the other finishing teams in the plant might benefit from what they have learned. Therefore, the other finishing teams took a short tour of the new bookcase finishing area and then had a meeting with the problem-solving team to see the results of its new methods, encouraging the formation of other problem-solving teams in the plant. The new teams consulted a refinishing expert who showed them how lacquer thinner could be used to melt and repair finishes. Such a technique allowed the crews to quickly repair bookcases that might have been scrapped.

OTHER TYPES OF ATTRIBUTE CHARTS

The Xeron team had two other types of attribute control charts from which to choose. The first was the np-bar chart for those situations requiring the monitoring of defectives using a constant sample size. The other was the u-bar chart for determining the average number of defects per unit of production.

The np-bar chart. The np-bar chart, like the p-bar chart, monitors defectives. However, the np-bar chart requires constant sample sizes because it records the number of defectives found (the so-called "np"). Since the sample size is constant, control limits also remain constant.

The Xeron management team could have easily employed an np-chart in the planing area of the bookcase line. Workers in this area smooth the rough boards before they are cut to the proper length in the cutting room. The planers work only with groups of 100 boards at a time since this is the number delivered on each skid. Therefore, the team could be told to monitor for defective boards caused by cracks, gouges, chips and other such defects. The number of defective boards found in each bundle of 100 could then be tallied on the chart.

An np-chart could be created following the same steps as for any attribute chart. The process potential study would require 25 groups of 100 boards to be examined by the team using an established defect checklist. The control limits are then calculated using the formulas listed below. The final chart is posted for at least 30 days and statistical control is demonstrated. Then the team would work to reduce the total number of defectives produced in the planing room.

The u-bar chart. The u-bar chart is an alternative to the c-bar chart. The u-bar chart tracks the number of defects per unit. The Xeron company's bookcase line has another concern that a u-bar chart could monitor. Some of the finished bookcases are later equipped with brass corners and sliding glass doors in the final assembly area. The assemblers are aware of a host of annoying defects at this stage that mar the beauty of the final product. To monitor the number and types of defects they could use a u-bar chart. Sample sizes can vary with a u-bar chart, so the assemblers would be able to inspect as many hardware pieces as possible during SPC charting breaks. The number and type of defects could then be tallied on the chart. The number of defects per unit would be plotted on the chart. The u-bar chart is used like a c-bar chart to find and eliminate the cause of defects. For example, they may find that the brass pieces dent and scratch easily. The suppliers could be requested to cover the brass pieces with a peel-away layer of plastic to help prevent handling damage.

Number Sampled	Number Found Defective
50	12
50	18
50	22
50	18
50	22
50	21
50	23
50	15
50	24
50	13
50	14
50	18
50	23
50	30
50	17
50	15
50	14
50	17
50	20
50	19
50	15
50	15
50	12
50	22
50	17

Average Sample Size   50   Total   456

$$k = 25 \ (\text{ number of subgroups })$$

$$\frac{456}{25} = 18.24 \quad \text{or, the average number of defects } (n\bar{p})$$

$$\text{UCL / LCL} = n\bar{p} \pm 3 \times \sqrt{n\bar{p} \times \left(1 - \frac{n\bar{p}}{n}\right)}$$

$$= 18.24 \pm 3 \times \sqrt{18.24 \times \left(1 - \frac{18.24}{50}\right)}$$

$$= 18.24 \pm 3 \times 3.40$$

$$= 18.24 \pm 10.2 \qquad \begin{array}{l} \text{Upper Control Limit} = 28.44 \\ \text{Lower Control Limit} = 8.04 \end{array}$$

Figure 3.11 The calculation of control limits for an np-chart.

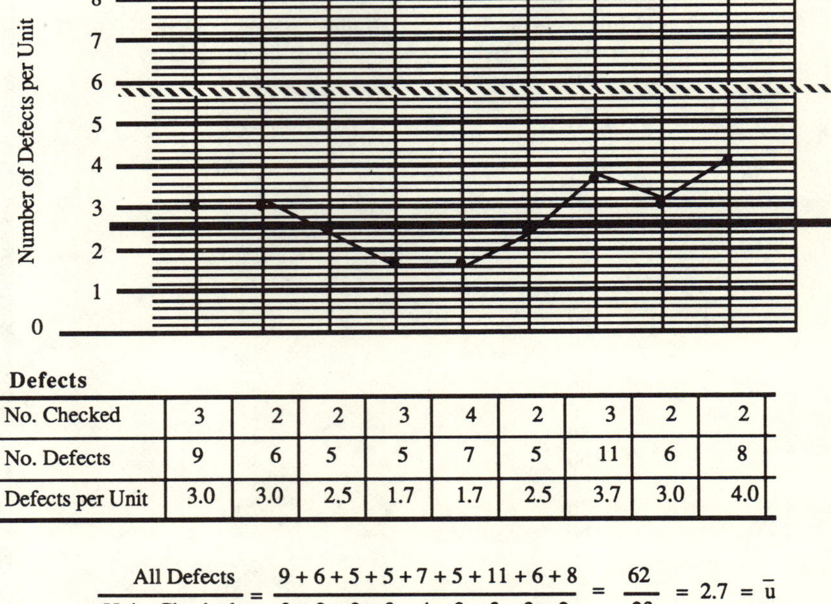

Defects

No. Checked	3	2	2	3	4	2	3	2	2
No. Defects	9	6	5	5	7	5	11	6	8
Defects per Unit	3.0	3.0	2.5	1.7	1.7	2.5	3.7	3.0	4.0

$$\frac{\text{All Defects}}{\text{Units Checked}} = \frac{9+6+5+5+7+5+11+6+8}{3+2+2+3+4+2+3+2+2} = \frac{62}{23} = 2.7 = \bar{u}$$

$$\text{UCL} / \text{LCL} = \bar{u} \pm 3 \times \sqrt{\frac{\bar{u}}{n}}$$

$$\frac{23}{9} = 2.55$$

Average Sample Size

$$= 2.7 \pm 3 \times \sqrt{\frac{2.7}{2.55}}$$

$$= 2.7 \pm 3.09$$

Upper Control Limit = 5.79
Lower Control Limit = -0.39, or zero

Figure 3.12 An example of the calculation of control limits for an np-chart. Normally 20-25 subgroups of data are used for this calculation.

Special Note: Chart Failure at Low Levels of Defects. Attribute charts are unique because they have a level where further improvement is impossible. That level is, of course, zero. Zero defects, defectives or zero percent defectives are as low as plotted points can go. In fact, when a chart, such as a p-bar, has a baseline below 5%, the statistical assumption that made the chart possible begins to fail. In other words, any attribute chart that begins to register frequent zero readings is losing statistical validity, meaning that the control limits and the plotting are now valueless as information. However, it does not mean the control chart is useless. On the contrary, only the method of using the chart changes.

First, the charting area of the chart is no longer used, unless customers specifically require it be filled out. Second, the defect tally area becomes the focus of attention. When low levels of defects are occurring, then there will be periods of zero defects. Two or three of these periods in a row indicate the process is capable of producing zero defects for short periods of time. The idea is to stretch these periods out. This is done by instructing the operators to react to the detection of any defects. Even if only one minor defect suddenly is found in the process a reaction is still necessary. If the process were not producing that defect for several hours, why did it start to do so now?

Some have argued there is a point of diminishing return on the elimination of defects i.e., eventually the cost of eliminating the last few minor defects will be more expensive than the cost reduction for production. However, the intangible costs of defect elimination can usually counter this argument. Once zero defects are achieved, then the production process can assure perfect service to the next stage of production. For example, if the machine that makes bolts to be used in the assembly area produces defect free bolts, then the assembly people can adopt a Just-in-Time (JIT) delivery schedule for bolts with confidence. Without zero defects, a JIT system would run afoul of stoppages to replace defective goods.

USE OF DEFECTIVE CHARTS FOR IN-PROCESS CONTROL

A major concern for some manufacturers is the value of adopting the defective charts for in-process control. Concern exists because the defective charts require samples of at least 50 for accurate monitoring of a process. However, there are p-bar and np-bar charts being used to sample 10 parts an hour. The usual application found in many factories is a process with a go/no-go gage used to determine whether parts will be rejected. The control chart monitors how many of 10 parts every hour are rejected. Usually a reaction is evoked if any of the parts fail.

The problem is that such an arrangement would allow high levels of defectives to slip through undetected just by chance. The other problem is that the go/no-go gages are usually set to the upper and lower tolerance

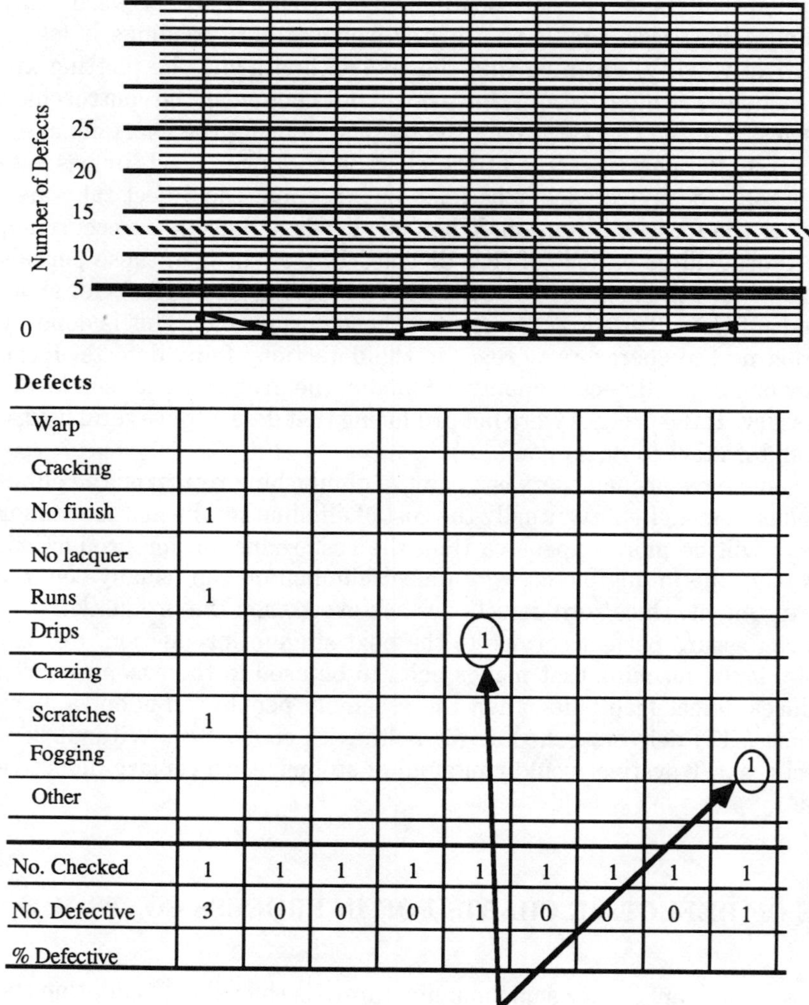

Defects									
Warp									
Cracking									
No finish	1								
No lacquer									
Runs	1								
Drips						1			
Crazing									
Scratches	1								
Fogging									1
Other									
No. Checked	1	1	1	1	1	1	1	1	1
No. Defective	3	0	0	0	1	0	0	0	1
% Defective									

When the level of defects falls to the point were most readings are zero, any occurence of a defect is circled and treated as a point out-of-statistical-control. Note how control limits become useless at this point.

Figure 3.13 The use of an attribute chart at low levels of defects.

limits for a dimension. This does not promote the "tighter-than- tolerance" requirements that exist today.

The proper solution for this situation is to use an X-bar/R chart and to have the operator equipped with the correct measuring device. The variable information from revision of the system would provide far more valuable information while requiring a smaller sample size.

A quality assurance manager in Michigan was once asked what he was going to do with his company's go/no-go gages after its new SPC system was complete. "Paint them grey and throw them in the lake," was his reply.

SUMMARY

For a direct connection between monitoring a process and evoking problem-solving methods, nothing compares to an attribute chart. Many of them are based on concepts that the work force and management already understand, such as scrap rates and defects per unit. By graphically portraying this information and training people that reductions are the only trends acceptable on a chart, great strides in improvement are possible. Of course, progress is not possible unless the charts are supported and maintained. In addition, there must be more than just control charts on a production floor for companies to realize a return on their investment in SPC. The next chapter will detail the complete SPC system and its requirements.

CHAPTER 4
THE COMPLETE SPC SYSTEM

The previous chapters have shown how a control chart can confirm that a process is stable, predictable, and capable. The remainder of this book will address the problems of moving beyond SPC to the goals of achieving higher profits, more efficient operations, and a happier work force. In other words, what lies beyond control charts to make the system complete.

A true story from a manufacturer will illustrate why more than control charts are needed for a complete SPC system. A company manufactured several kinds of products, one of which involved the welding together of two pieces of steel strapping. These straps were about a foot long. The customer specified that the weld must withstand a pulling force of at least 300 foot-pounds. The company tested the welds by sending them to a testing lab located in another part of the plant. The lab technicians mounted a few sample pieces on a machine that pulled them until 300 foot-pounds were achieved, or the piece failed. The production department was notified if any piece failed. One day, a new lab technician was hired who was instructed to test to 300 foot-pounds, but his curiosity got the better of him. Instead, he tested until the piece failed. He also charted the energy at which the piece failed and calculated that the average weld he was testing failed at about 500 foot-pounds.

Several weeks later a strange thing happened. The lab technician was testing the sample pieces that had accumulated for two days and discovered they failed at an average of 1200 foot-pounds. This was more that twice the strength of previous pieces. Obviously, the production people had discovered a better welding method. The technician took the results straight to the plant manager to announce the good news. The plant manager was impressed by the improvement and called in the welding foreman. The welding foreman was delighted, but was at a loss to explain the new method, since he wasn't aware of one. The three men went to the welding area to ask the operators what the new method was. The operators were confused because they too were unaware of any new method. The plant manager asked, "What did you do differently for the past two days that made these welds stronger?"

"Nothing we can think of", replied the operators.

"Let's test today's production", suggested the foreman.

The results showed that the pieces failed at an average of 500 foot-pounds. It was a mystery how the welds could increase in strength for two days and then go back to their normal levels. Or was it?

A complete SPC system is characterized by a company's work force knowing of the pressing need to monitor the capabilities of all processes and learning ways to control the processes. The above example illustrates this point. First, something did happen to more than double the strength of the weld, but, no one was looking for the improvement because there was no indication from management that this was a priority. Second, it is no surprise that people can't remember anything different happening on the shop floor. Most individuals can't remember what they ate for lunch two days ago. Lunch, like some jobs, is so routine people tend to disregard slight changes. Finally, where are the tools to help find the changes that improve the process? The increased weld strength was a potential improvement that was lost because no one was looking for it. The testing discovered the improvement too late and, there were no records of past activities.

THE COMPLETE SYSTEM

The intention of a complete SPC system is to present a company's single operating priority to the employees along with the tools and support to achieve a related list of goals and objectives. The single priority is also called the company's *operating philosophy*. Every company has an operating philosophy whether someone has stated it or not. In a complete SPC system, the philosophy has been well thought out and clearly stated by the management. Part of a well-designed philosophy is a list of goals to be achieved in the near future. These goals and a schedule of objectives, are called the company's *strategic plan*. The strategic plan is the blueprint for how the company plans to remain competitive.

The following is a list of the steps that are taken to create an SPC system for the increased competitiveness of a company:

1. Management Commitment and Writing the Strategic Plan
2. Forming the SPC Committee
3. Developing the Quality Plan
4. Training the Work Force

5. Gage Control, Print Control, and the Process Log
6. Capability Studies
7. Control Charts
8. Reaction Plans
9. Problem-Solving
10. Continuous Improvement
11. Involvement of Suppliers and Subcontractors
12. Increased Competition
13. Improved Design
14. New Methods (Just-in-Time, Zero Inventories, CIM)

MANAGEMENT COMMITMENT AND WRITING THE STRATEGIC PLAN

If the top management of a company has not made a commitment in both writing and resources to the implementation of an SPC system, there is no need to proceed further. The SPC system will fail. Every critical phase of implementation relies on top management settling disagreements or providing the support needed to get the job done. In addition, all of the top managers must agree on the direction the company chooses to pursue. If two opinions of how to implement SPC should form, then the company will probably split along these lines as well. The strategic plan can both illustrate and prevent these problems.

The first step for top management is to make a commitment to implement an SPC system by publishing the operating philosophy for the company. The operating philosophy is a policy statement about why the company is in business. Typically, this operating philosophy says something like, "Our purpose is to meet or exceed customer requirements and desires." A company can have only one operating priority in its philosophy. A company can only go in one direction at a time. This is why a lot of companies have very general operating policies. Ford Motor Company summarizes its policy in the advertising phrase, "Quality is Job 1." Such a policy must be widely published and discussed so that all employees and customers clearly understand the company's position.

The strategic plan puts teeth into the operating philosophy and is the guideline for the development and change of the company over the next several years. The typical strategic plan covers a period of five years. The plan is developed from the operating philosophy. Xeron's operating philosophy might be that "Xeron should produce the highest quality products for customers and best quality environment for its workers." Now, top management of Xeron must meet and decide what goals, objectives and timeliness will be involved in implementing this philosophy.

For example, "zero defects" could be a major goal of such a philosophy. Management must define this goal in writing, state how it will be achieved, note which objectives will be reached on the way to this goal, and suggest when these objectives should be reached. The results may look like this.

- Goal: Zero defects in production parts.
  To maintain Xeron's philosophy of producing the highest quality production parts, all product defects must be removed. This will be measured in the final inspection reports.
- Objective: 50% reduction in reported defects by June of next year.
  Production will implement a series of defect control charts at selected stages of production to document the number and causes of defects. Quality control and production will work together to take corrective actions that will reduce the number of defects reported by final inspection.
- Objective: The creation of problem-solving teams.
  Personnel will be in charge of training production teams in troubleshooting methods and problem-solving techniques. This training must be completed by June of next year.
- Objective: 90% Reduction of defects.
  Industrial engineering will be responsible for collecting the information collected by production and quality control. They will use this information for the design of new processes and methods that prevent the defects. The effects of these changes will be reviewed by the SPC committee. Objective should be reached within two years.

The objectives relate to the goal. The intention behind each objective is briefly noted and the objectives make somebody responsible for the successful completion of the action. These are the signs of a good strategic plan. It must be specific enough so people have the responsibilities, yet flexible enough to allow for change because of unforeseen problems. Top management must create this strategic plan and then publish it so all other managers know what actions they are expected to take during the next few years.

There are a few guidelines to follow when making a strategic plan. First and foremost, all disagreements should be addressed during the creation of the plan. Managers who don't support the plan later will only slow implementation. Disputes should be settled and compromises reached in the early phase of planning. Second, the main concern is not making the plan perfect. The plan should reflect desires—not be a rigid set of rules. This includes a section on how to revise the plan to fit current needs. Such changes should have the signatures of the top managers. Third, detailed descriptions of actions should be restricted to the first year of the plan. Each year's goal will be elaborated upon after information becomes available from the production process. Finally, progress should be demonstrated with facts, not opinions. The measures of success must be established before SPC implementation. In the above example, the final inspection reports will be the measure of success. Avoid financial measures where accounting practices can easily alter the perception of the actual situation.

FORMING THE SPC COMMITTEE

One of the goals of the strategic plan will be to implement a system of statistical process control. This is best accomplished when an SPC committee is appointed by the chief executive officer (CEO). The CEO must not be a member of this committee. The role of the CEO is to appear in front of this committee from time to time with a copy of the strategic plan in hand. The CEO should ask questions such as, "Where are we with the development of problem-solving teams?" or "Why aren't the scrap rates down by half as they should have been a month ago?" In other words, the role of the CEO is to be the driving force behind the implementation of the strategic plan. Groups like the SPC committee should be taking the steps needed to implement the actions specified in the strategic plan.

An SPC committee can be from one to nine members in size. More than nine members tends to rob each participant of the feeling of playing a vital part in implementation. The membership of the committee should be composed of the minimum number of people vital to implementation. For example, Xeron's committee would be made up of the top managers from each department. Small companies usually use the quality control manager, the production manager or both, as the SPC committee.

The committee votes on what actions will be taken by whom, where, and when. Then the manager for the affected area must take the action and report on its progress to the committee. The first meeting of the committee involves the review of the strategic plan. Each goal is discussed and a vote is taken as to which ones to work on and which ones should wait. The Xeron committee decides to implement the zero defect goal first. In discussion, the Xeron committee decides that the shipping area should be included in the no defect goal. In other words, "no defects in the products" also means that no errors are allowed in shipping orders. The manager of material handling is instructed to construct a system for monitoring shipping errors and to report on this system to the committee in one month. A month later the system is reviewed by the committee and approved. At that meeting, the material handling manager is told to implement the system and to report on the number and types of errors in another month. The material handling manager then calls together the material handling supervisors the following day and instructs them in the use of the system. The supervisors in turn instruct their personnel. Information would be gathered and summarized for the supervisors. The material handling manager would carry this information back to the committee. The committee would then vote on a schedule for improvements and insist that the material handling manager report on the progress made each month.

This process forces actions to be taken after every meeting. The implementation must be continuous if SPC is ever to become a reality in Xeron. For problems with nonproductive meetings, there are several excellent guides available about effective meeting methods. The SPC committee must be trained in efficient meeting techniques. This way, the efficient and professional meeting methods will inspire imitation by other meeting groups.

As a final note, nothing should be secret about an SPC committee. To help implementation, the committee should regularly report goals and results to the rest of the company. The company newspaper or bulletin board is a good place for summaries of actions, successes, and problems encountered. Participation is encouraged by allowing people to submit suggestions so that make all employees feel that they have a part in the process.

DEVELOPING THE QUALITY PLAN

One of the most popular themes in today's corporate philosophies is the emphasis on quality. Quality of the product, quality of work relations, and quality of services provided are common. Therefore, the SPC committee must address the theme of quality by developing a corporate *quality philosophy* and related *quality plan*. The quality plan is a strategic outline of how the company plans to study and gain control of the quality of production. One of the main tools to be used will be SPC. In addition, the quality plan will detail how inspection procedures will be developed and used. The intention is to create a formal system for the measurement of quality. For example, what will be the standards for judging the severity of defects and how rigidly will these standards be enforced? Without a clearly defined system, people will tend to change judgment rules to fit each situation, instead of striving for real improvements.

TRAINING THE WORK FORCE

The most important consideration for the implementation of SPC is training which should provide all affected employees with the skills necessary for starting and developing an SPC system. Training can be performed by outside organizations, internal personnel, or combinations of both. Whichever method is used, it is vital to prepare several items before contacting the instructors. The most important of these is to write out a list of skills participants should possess after the training. Second, there must be an established method of evaluating training effectiveness. Third, people who attend the training should be able to put their new skills to work as soon as possible. If operators are trained in the use of control charts, then control charts should be in place at the work stations before the training is completed. Otherwise, within a month or two the newly acquired skills could be lost from lack of practice. Whenever training takes place, a quiet and comfortable location for instruction is a must. The training room should include a blackboard, a flip chart of blank paper, and any other required audio-visual equipment. Finally, the number of people who need to be trained has to be determined as well as times they would be available for instruction. This is published as a training schedule and must be accompanied by a short memo explaining the purpose and importance of the instruction.

When negotiating for training, a company must have proven performers who will adapt their program to the companys needs. These needs are far more important that the desires of the instructor. If a local college wants to conduct a regular class in SPC for college credit, the company should request that a program be designed around its own processes. If the college claims it has proof of the effectiveness of its method, then the company should reconsider. If on the other hands, the college only uses a particular format to get state reimbursements, the company should remind the college of who's paying for the course. The same principles apply to business consultants or independent training groups. The company should require evidence of the success of the program and references from similar companies who have benefited from the training.

When training is conducted, the training coordinator and at least one member of the SPC committee must be present. The training coordinator is there to confirm that the standards for training are being met and that any requests from the participants are acted upon. For example, if a dispute develops about whether smoking is allowed in the training room, the coordinator can help settle the matter. When possible, the CEO should start each session with a two minute review of the strategic goals and why this training is important. The presence of the CEO or other top manager at the first session tells the employees this training is important.

SPC training is conducted in several forms. Basic techniques of SPC are usually given to all employees in a short set of seminars lasting no more than ten hours. Extensive studies of the SPC methods are typically given to the managers and other key people in the SPC system. Inspectors get something more than basic SPC training. The CEO and staff are trained in the use of SPC and the role of management in the system. Currently, companies such as General Motors, Ford, and AT&T have classes for CEOs. Attending these sessions allows one CEO to mingle with other company executives who are implementing SPC.

After the basic SPC system is established, additional training is recommended. This type of training can range from an Advanced SPC class to reviews of basic machining principles or the use of microcomputers to automate the system. These topics should be selected according to the needs of each industry.

GAGE CONTROL, PRINT CONTROL, AND THE PROCESS LOG

These three parts of an SPC system are fundamental to establishing control over production processes. The process log is the most important of these three parts. The process log is an hour-by-hour record of what is happening at each stage of production. Ever time an adjustment is made such as, a new material added, a new operator on-site, or any other change in the situation, it is recorded in the process log. This log is no different from those used by a ship's captain. Each change in the condition of the ship or its course is recorded. If a problem arises or the ship becomes lost, it is easy to retrace and correct the causes. Even if the ship sinks, the recovered log helps other ships from suffering the same fate. An industrial process is no different from a ship at sea; it too must keep track of every change, or risk becoming lost.

A previous example demonstrated how an increase in welding strength came and went without anyone being aware of its cause. Proper documentation of production operation can prevent this problem. An example is the machine operator at an FDA regulated company. One of the regulations from the FDA is lot traceability—that is the ability to trace all of the materials that went into making a product. For example, with instant ice tea mix, an FDA inspector might come to the company with a jar of ice tea purchased in Des Moines, place the jar on a table and say, "Show me all the records of all the materials that went into making this particular jar of tea." Tough regulations, but the traceability is invaluable when contamination is detected in a product.

Much of SPC will rely on accurate measurements and observations. Therefore, before SPC is implemented there should be a formal program of gage control. Gage control involves three parts. First, the physical control of the gages must be regulated, and they should be numbered and checked in-and-out with each shift. Second, each gage must be calibrated at least once a year; usually a small sticker is placed on the gage indicating the next time it is due for calibration. Finally, gages have to be tested for repeatability and reproducibility. Simply put, repeatability is the amount of error when a gage repeatedly measures the same dimension on the same part. Reproducibility is the error created when different people use the same gage. The combined error is called the R&R of the gage. The R&R should not be more than 10% of the tolerance spread for the measured dimension. To keep this system as simple as possible, each part or job is assigned a specific set of gages (see Chapter 9).

Blueprint control is needed to confirm that only the most recent prints are being used on the shop floor. The system is traditionally created and maintained by the engineering staff. As updated copies of a blueprint arrive the existing copies are collected and destroyed. No stray prints are allowed.

CAPABILITY STUDIES

At this point an SPC coordinator should be identified. The SPC coordinator is in charge of systematically implementing control charts and later summarizing the information from these charts. The SPC coordinator is responsible for conducting capability studies on each stage of production. The actual data collection can be left up to the operators at that time or to trained technicians. Only in the smallest of plants should the coordinator collect data. The coordinator's time is better spent integrating control charts into the production process. Capability studies are used to create these control charts. Also, the SPC coordinator maintains a list of all control charts on the production floor and their current status. In short the SPC Coordinator is the resource person for the immediate operation of the SPC system.

CONTROL CHARTS

Once a control chart is placed on the floor, the coordinator should meet with the supervisor to review the necessity for keeping timely information on the chart. In addition, the coordinator checks each chart at least once a week to ensure that proper procedures are being followed. See Chapters 2 and 3 for more complete descriptions of the creation and use of a control chart.

The control chart is never alone on the production floor. Usually an "SPC Packet" is assembled so the operator has all relevant material in one location. The operator is responsible for filling out the control chart. A typical SPC Packet will contain:

1. The control chart for this particular part.
2. The process potential study that explains how well the produced part should conform to specification.
3. A set of blueprints for the part.
4. A copy of the gage study with R&R clearly denoted.
5. The process log for recording all changes made in the process.
6. A reaction plan for out-of-statistical-control conditions.
7. A control chart log for recording actions taken to correct problems detected by the control chart. This log is usually located on the back of the control chart.

In addition, these packets should be contained in a plastic cover to protect them from the factory environment. Nearby should be the proper gage and a suitable work area for the above material. Finally, production scheduling is modified to allow operators time to use the SPC system.

REACTION PLANS

A control chart is just another piece of paper unless a reaction plan is included, which contains the steps an operator takes whenever the control chart indicates that the process is out-of-statistical-control. The reaction plan is also an ever-changing document. The first reaction plan is a simple set of instructions, such as, "When loss of control occurs, stop the process and notify the supervisor." After several corrections are recorded on the back of the control chart, the SPC coordinator can review these to update the reaction plan. For example, a grinding process's control chart log might show that most corrections involve adjusting the grinding wheels. Therefore, the reaction plan is updated to include a table

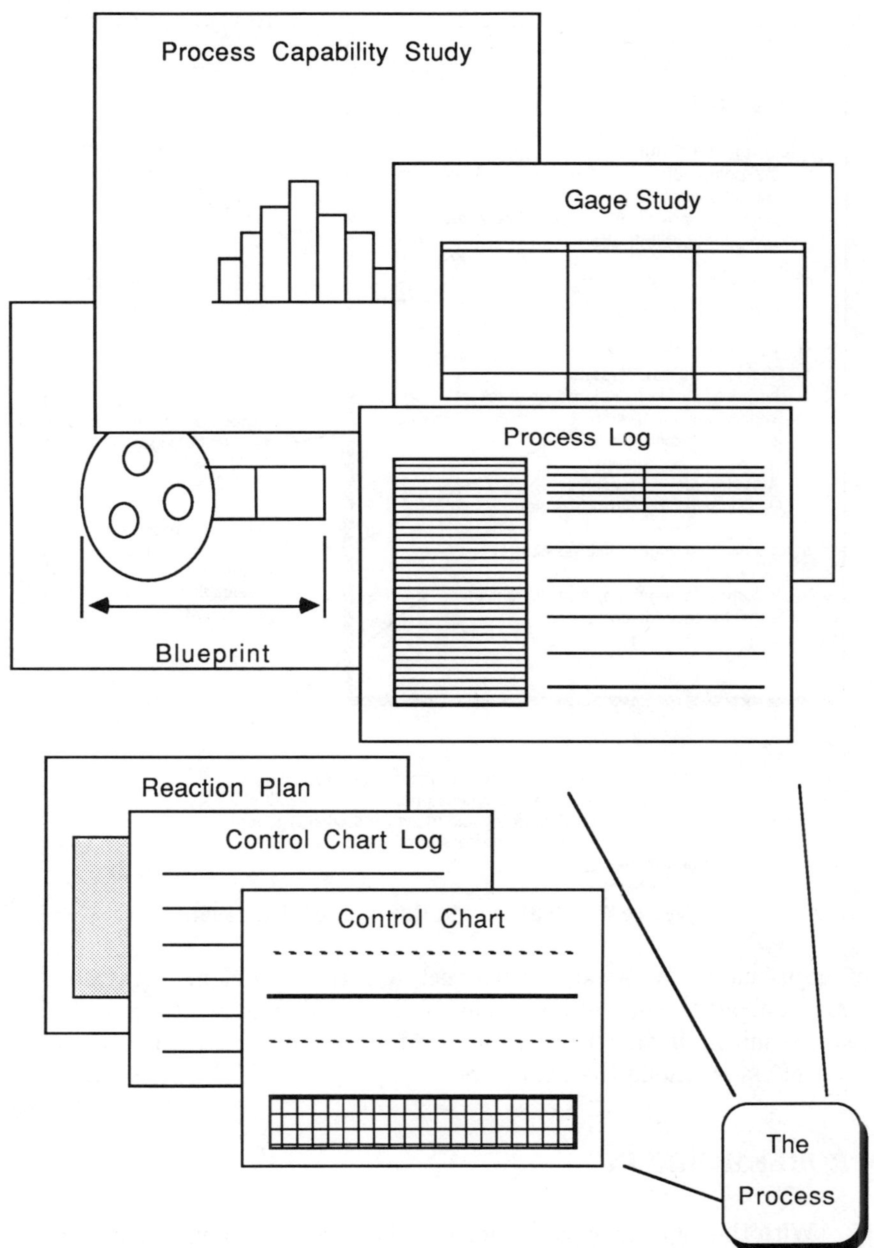

Figure 4.1 The key elements of an SPC system.

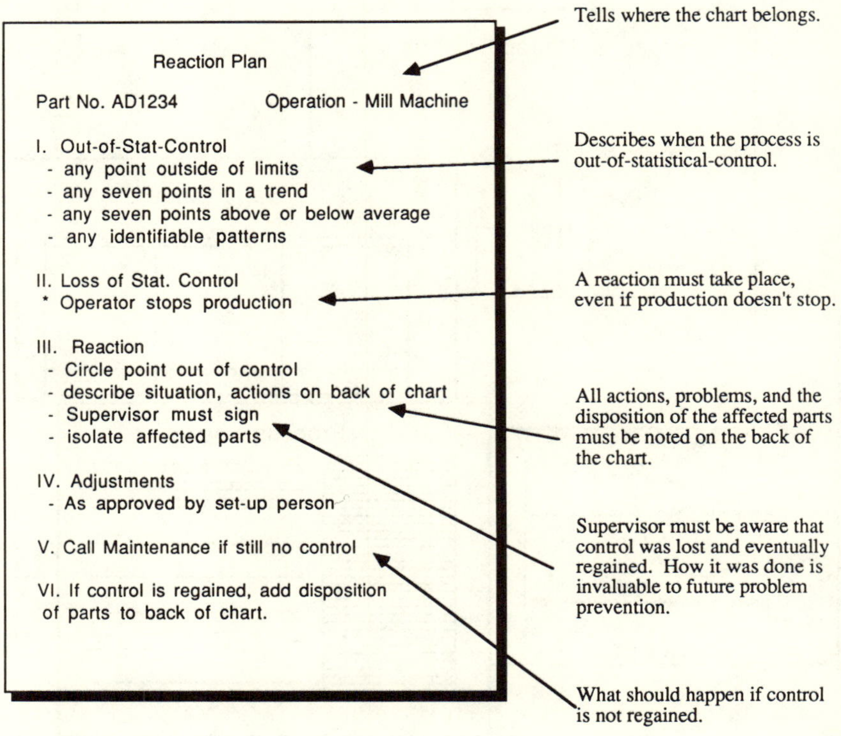

When reactions are documented, the rule of thumb is that a person who has never seen this process would be able to read the control chart log and understand what type of actions were taken. Supervisors should initial all noted reactions.

Figure 4.2 Key elements of a reaction plan.

of adjustments necessary to correct different problems. As more is learned about the process, more of this knowledge goes into the reaction plan, resulting in shorter lead times, effective corrections, and a higher level of skills among the operators.

PROBLEM-SOLVING

With the capability study done, a chart in place, a reaction plan for the chart, and gages calibrated, a physical SPC system exists. Now all efforts focus on using this system to increase the competitive position of the company. Since the SPC system will be documenting all sorts of problems, the first step is to introduce mechanisms for problem solving. The most effective medium for these problem-solving mechanisms is the work team.

History has shown that problems are best solved by groups of people. Even those individuals who made great discoveries had small groups assisting them. In the manufacturing environment, the work team may already exist as small groups of operators working together on similar machines. Management needs to train and encourage these groups to work together in solving common problems. Such groups have a variety of names such as, Quality Circles, Employee Involvement Teams, Quality of Work Life Teams, or Process Improvement Teams. The name is not important; the mission is.

Top management must create opportunities for these teams to make contributions to the improvement of the production processes. Consequently, time must be set aside at regular intervals for the teams to meet and discuss production problems. A representative of each team should be sent to the SPC Committee meeting to secure permission to tackle a specific problem. The SPC Committee should keep a list of priority problems that the teams may refer to for selecting projects. Teamwork training and a good group leader are essential to a successful team effort. Usually, membership in the team is voluntary to keep the participatory spirit alive. Successes from a team are widely published to encourage the formation of more teams (see Chapter 11).

When a team is selecting, working on, or evaluating a problem, control charts provide invaluable information. The baseline of the control chart can be used to measure the success of the team effort in improving a process. Attribute charts can provide a list of the most frequently encountered defects on the production floor, thus making project selection easier when using methods such as Pareto Analysis. For example, a control chart for scrap rates could help guide a team trying to improve productivity. Those ideas that reduce scrap rates will be detected by the chart.

CONTINUOUS IMPROVEMENT

It is the responsibility of management to seek new ways to improve the competitive position of the company. This is not a one-time goal, but a continuous process. Essentially, continuous improvement is the operating philosophy of any competent management. However, a complete SPC system requires this philosophy to be part of all operations within a company. Any process can always be improved. It can be made to run faster, operate at lower costs, or achieve higher accuracy. As has been pointed out by Dr. Deming and others, only by seeking such improvements can a company hope to maintain its market share in today's world economy (see chapters 1 and 11).

The work teams and the rest of the SPC system must provide management with detailed information about the capabilities of the manufacturing processes. With this information, management can make decisions about which new capital resources are needed, when maintenance should be increased, or when some products should be discontinued. For example, management of Xeron corporation may observe that several teams have had problems obtaining information about tool history. Therefore, they may decide to install a computer database that tracks tool histories and makes these available on request.

INVOLVEMENT OF SUPPLIERS AND SUBCONTRACTORS

Many of the companies that have implemented SPC have discovered that the cause of some of their problems comes from outside the company. Specifically, they find that the quality of materials from suppliers or the work of subcontractors varies too much. Therefore, implementing SPC should also include these support organizations.

Suppliers are the easiest to involve. First, they are keenly interested in keeping business, and second, most companies already monitor the quality of supplied materials. The purchasing contract is the best method for encouraging an SPC system for a supplier. Ford Motor Company states plainly if a company doesn't have an SPC system, then Ford is not interested in its product. This is not a large corporation browbeating small firms. The information gathered during the production of these parts is invaluable to Ford in locating and eliminating production problems in their own plants (see chapter 14).

An illustration will emphasize this point. A fictitious company that assembles steel shelving units has subcontracted to another company to weld nuts onto the inside of the shelves. Specifications require that these nuts should be able to withstand a torque force of 60 foot-pounds. In the first company's shop, bolts are driven through the nuts with a torque wrench and the torque wrench's force is recorded on a control chart. It would be very beneficial to receive welding test results from the subcontracted company with every shipment. That way, if the welds tested broke at 60 foot-pounds, as required, the operators of the torque wrench could be alerted that torque force on the wrench must be held between 50 and 55 foot-pounds for this lot of parts.

By far, the best way to involve vendors and subcontractors with an SPC system is to form a working partnership. Management of subcontracting companies should be brought to the plant and shown the SPC system. Emphasis should be on the many benefits realized from this system. In the above illustration, the supplier of the welded nuts should be interested in the results of assembling components. Any information about what makes the shelves assemble better will benefit a supplier interested in improving productivity.

INCREASED COMPETITION

Once an SPC system is implemented and producing success stories from the work teams, then the system can become more competitive. Top management, sales, and purchasing can use the results to increase the market share of the company.

Top Management can use the information from the SPC system in decision making for new equipment or methods. A metal stamping plant may discover that half of their dies are not capable of holding customer specifications. By consulting with the cost accountant, they can decide which dies are cost effective to fix or replace. If information indicates that production with one die will produce some scrap, then management can decide if it is still profitable to sort the production or to cancel the job.

Having documentation that the production processes are stable, predictable, and capable is a powerful sales tool. The maker of an ashtray for automobiles once came into a sales presentation with 182 days of control charts showing that key specifications were held to a variation of less than one-half of the customer's tolerance range. Needless to say, this company retained the current contract and won several more. Most customers should be willing to pay a higher price for a product they know will do the job. Of course, the surprising thing about SPC is that the price should be lower for a predictable product. SPC should have reduced the scrap and defect rates in production. Less rework and scrap means lower production costs, thus a more competitive price.

Purchasing plays a key role in increasing the competitive edge of the company and is responsible for obtaining materials that will work effectively the first time. If this is accomplished, the cost of production is reduced. If material were defective at the time of purchase, then the production people are wasting time and money trying to make it into a quality product. For instance, a steel stamping operation discovered from its control charts that every time a new roll of steel was fed into the machine, the dimensions of the part changed dramatically. The operators measured the thickness of each roll of steel and found they varied widely from roll to roll. The operators also knew from experience that if the thickness of the steel changes, the size of the stamped part changes too. They took this information to the SPC committee on a day when the CEO was visiting. The CEO asked why the purchasing manager didn't buy steel that was certified at a specific thickness. The purchasing manager replied, "Because it cost fifty cents more a ton."

The CEO said, "You know, I bet it would be cheaper to pay that fifty cents now than to pay for the cost of sorting parts later."

The CEO gets an A+ in economics.

IMPROVED DESIGN

By this time, many people ask where the Japanese are now with SPC. As it turns out, one would be hard pressed to find a control chart in Japan today. The Japanese have studied their processes for forty years and have advanced the concepts of SPC to the next logical step. That is, they have moved from "defect detection" to "defect prevention" and on to "defect elimination". This advanced philosophy of quality was pioneered by Dr. Taguchi. His philosophy is to design products and processes so the variation that will occur in production can no longer have an adverse affect on the final product. In short, to make the product "robust" against process variations (see Chapter 10).

Dr. Taguchi accomplishes this objective through the use of experimental design. In other words, products and processes are tested experimentally to determine the best design. To illustrate, if a plastic injection machine has 13 controls, what is the best combination of settings for a particular part? Setting each control to either high, medium, or low in various combinations unfortunately, would create about 1.6 million possible combinations. Experimental design is a method for finding the best combination using only a few representative combinations. Luckily, small computers make the calculation of the required statistical formula fast and painless (see chapter 13). The information can then pinpoint the best combination of settings.

Dr. Taguchi indicates that many companies are pursuing improvements in the design of products and processes through the use of Computer Aided Design (CAD) and Computer Aided Engineering (CAE). CAD allows the faster modification of existing blueprints. Since the Pontiac Fiero is drawn entirely with CAD, Pontiac could tell the computer to stretch the car six inches for better stability, test the increased stability with CAE, and have the machine draw up new sets of prints all in the same day (see chapter 9).

NEW METHODS

SPC does not promote the change of methods used by a company—simple economics does. However, SPC is required before new methods can be implemented. With CAD and CAE, the engineer can modify drawings faster, but what should be modified? The SPC system should feed engineers this type of information. If 60 degree bends in a copper pipe cause leakage problems, then new drawings should design out such inherent defects. If welders find that spot welding a part to a curved surface weakens the weld, then the engineers can simulate the loss on CAE to set new design specifications. In short, SPC and newer methods must work together for maximum efficiency. The philosophy of continuous improvement shouldn't stop with the control charts. SPC can be the doorway to other technologies.

SPC is rarely found alone on the agenda of new methods to implement. Usually, a company involved with SPC is also examining a Just-in-Time (JIT) production system. JIT is a philosophy of production that says that the ideal production process will have no excess inventory or production queued up. In other words, nothing is held just-in-case; everything is in constant motion. The idea of zero inventories is closely related to this philosophy. To accomplish JIT, the first step is SPC. How can the exact number of items to have on hand be planned unless accurate predictions of defectives or downtime are available? (see chapter 8).

Computer Integrated Manufacturing (CIM) and Flexible Manufacturing Systems (FMS) are current attempts at perfecting the fully automatic factory. However, automating a factory is not cost efficient. A worker in the third world making $1.50 an hour is far more competitive than a $150,000 robot doing the same job. The cost effectiveness comes from developing a plant that can modify the product or change jobs in seconds. A tall customer wants a desk that is two inches higher than a normal desk, places the order with the factory which modifies the design to produce one desk of this specific size. A few seconds later, the process is producing normal sized desks again.

To design such an automated factory an intimate knowledge of the processes of production is needed. Otherwise, automation will only magnify unsolved problems. SPC is vital to designing such a plant. The problems experienced by the conventional production process must be designed out of an automatic system. As an example, robots are marvelous at consistency, but if the material they are using is defective, they will continue to make defective assemblies until instructed to do otherwise.

HOW TO EFFECT CHANGE

The introduction of any new technology will be met by resistance. People resist change because they usually do not fully understand something new, and tend to fear what they don't understand. However, if the strategic plan for a company is balanced on a single, well-communicated theme, then implementation becomes easier. The idea of continuous improvement can give people a common vision. They can see control charts as just one part of a complete system that encourages competition through knowledge and improvement. Once the people are motivated, they supply the energies needed to reform the existing system to the requirements of SPC.

SUMMARY

SPC is considered both a method of management and a production control technique. SPC is fully implemented when problems are solved because of the information being gathered by work teams using control charts. After that point, continuous improvement is expected in each process, department, and management strategy. New methods and better techniques are recommended for any company that wants to stay competitive. The company that stands still only makes itself a better target for competitors.

As demonstrated by the discussion in this chapter, implementing a complete SPC system is a time consuming serious effort. Everyone is involved, from top management to all ranks of company personnel. Fundamental to this process is the existence of a strategic plan for implementing SPC and other technologies. The plan must be driven by a dedicated CEO who appoints committees of top managers and holds them responsible for implementation. Where unions exist, they too must share in the leadership role of implementing SPC.

Once, a consultant was asked how companies have managed to implement such a seemingly difficult task when their schedules are already so overcrowded.

"The same secret used to build the pyramids," he responded, "they tried real hard."

CHAPTER 5
PROBLEMS TO EXPECT WITH SPC— COUNTERMEASURES TO DEPLOY

The title of this chapter is a bit misleading. The chapter is not an examination of failures within an SPC system but shows how success can become the greatest threat to an SPC system. When first implemented, the problem-solving teams will probably find and fix large and expensive problems which will build the expectation that this will always happen. However, as more problems are solved there will be fewer large targets to attack. Enthusiasm for the program may decline as smaller successes are reported.

Most users of an SPC system rarely report that the control charts or statistical devices fail. Instead, their expectations are that various parts of the SPC system will work smoothly and integrate well the first time they are used. Most companies report that the implementation of a new system of management will involve a period of transition from the old methods to the new. During this time, many mistakes and problems may occur. Management of the company must impress upon employees that these problems have to be seen as learning opportunities. After all, the company is new to SPC and making mistakes is part of the process of finding the best way to match an SPC system to the needs of the company.

What follows is a brief list of the most common problems experienced with an SPC system and the steps that can be taken to prevent them.

PART 1: PROBLEMS WITH THE CONTROL CHARTS

It is rare for a freshly posted control chart to discover a process already under statistical control. Problems plague the chart, such as, mistakes made by the operators or difficulties because of the lack of the

Example of a chart in statistical control. The control limits have not
been exceeded, no identifiable patterns are present, and most plotted
points are near the process averages.

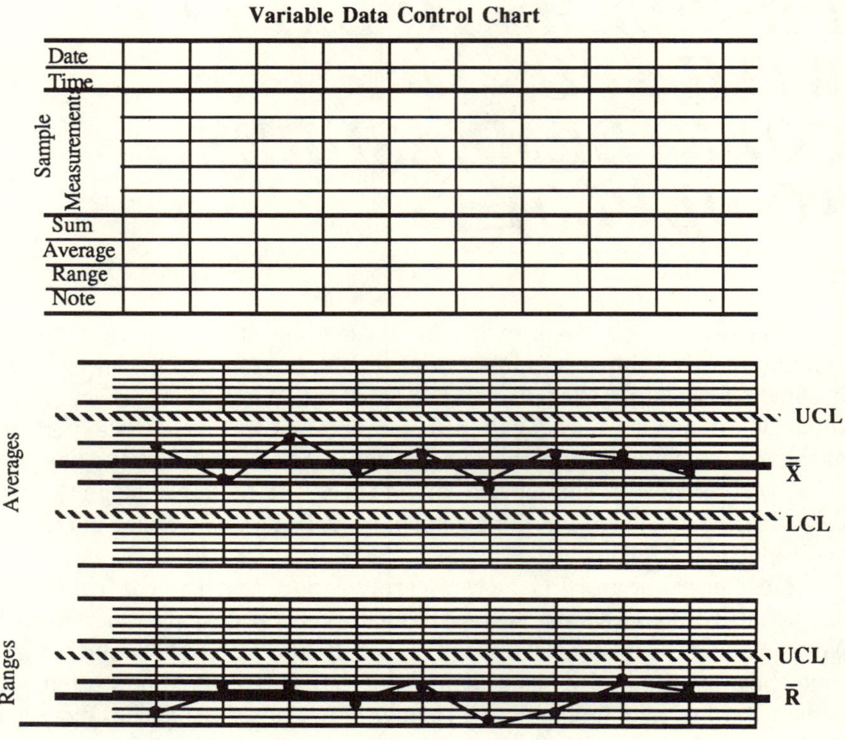

Figure 5.1 Example of a well behaved chart.

tools necessary for charting. These kinds of problems will create distinct
patterns on a chart. By being able to recognize these patterns and their
potential causes, a company can work quickly to correct the situation.

To begin, a company should be aware of what a process under statistical control looks like on a chart. The process under statistical control
will have points randomly plotted within the control limits, with two-
thirds of these points inside the middle third of the control zone. No
unusual or distinct patterns should be obvious. Once in a great while, a
point may cross a control limit, but corrective actions find the cause of
the deviation and quickly bring the process back into control. Such a point
is circled on the chart with the cause and solution duly noted on the back

of the chart. Since the control limits are based on 99.73% of the normal distribution, three out of a thousand points will cross control limits naturally.

Basically, anything differing from the random pattern above could be an indication of a problem. Identifiable problems and their possible causes are listed below:

Trends

The steady rise or fall of seven or more points in a row on a control chart is considered a trend. In the case of variable data, the cause is usually associated with the increasing or decreasing of the part size from tool wear, a fixture loosening, deterioration of material, or the normal performance of the process. Grinding machines have demonstrated the latter source during a "warm up" period.

On attribute charts, a trend upward is the sign of a worsening process, while a trend downward is the desired pattern for the chart. In the first case, reaction focuses on finding the cause of the situation and correcting it as soon as possible. In the latter case, the reaction is to find the reason for the solution to defects or defectives and to exploit it. If the trends persists, then new control limits must be calculated for the chart.

Two countermeasures are immediately available. The first is used when adjustments are possible. The beginning of a trend (seven points in a row) is stopped by adjusting the machine. The second method applies to machines that cannot be adjusted. The trend is studied to see if it is consistent. If it is, the control chart is tilted to match the trend. When a control limit approaches a specification limit, then the deteriorating tool or material is replaced.

The trend may be long term. If the saw operators at the Xeron company discovered that the saw blade slowly wore out over the course of 10,000 cuts, the more the blade wore out, the more variation they found in the cuts being made. By studying the control chart for the number of cuts made before quality was significantly affected, the team could predict when the blade would have to be changed. In this case, the blade was changed after 8,500 cuts. The team moved from detecting the problem with a control chart to preventing the problem using the information from the control chart.

One other form of a trend that appears mostly on attribute charts is the seasonal trend. For example, the Xeron company kept a p-bar chart on the scrap rates for the bookcase line. After the chart was kept for a year, the management team taped a year's worth of charts together. They discovered that from July to September, the percentage of scrap nearly

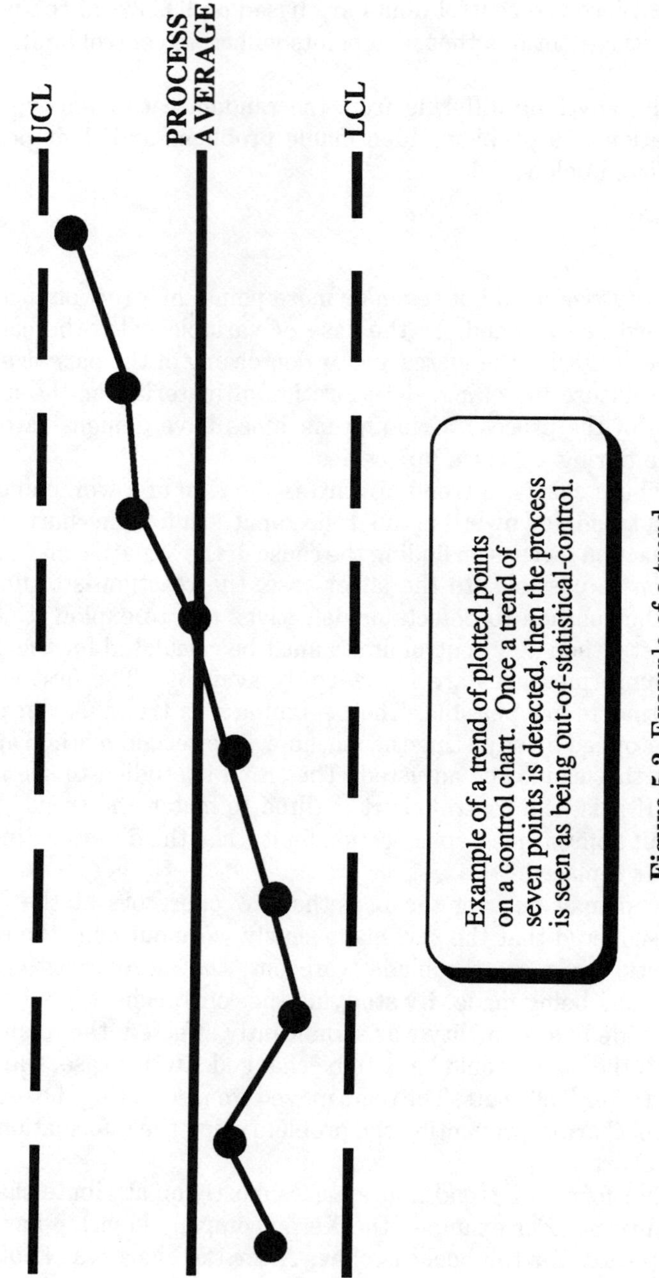

Figure 5.2 Example of a trend.

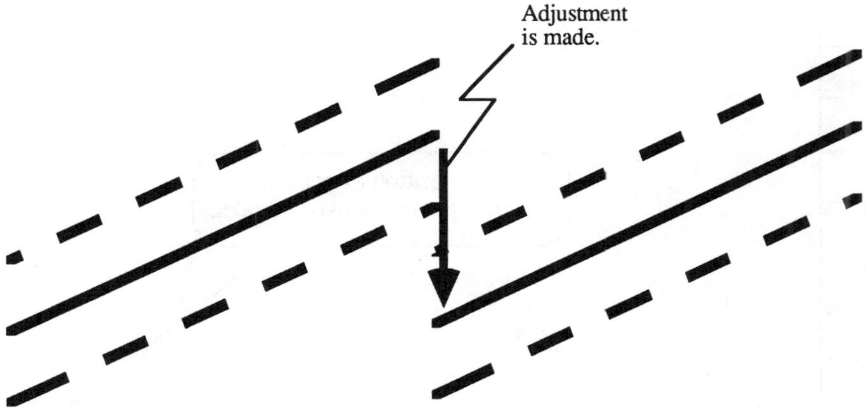

Once a trend is detected and found to be unavoidable (i.e. steady tool wear) then a regression chart can be used. The process average and control limits are tilted to match the rate of the trend. Once the value becomes too high, a machine or tool adjustment is made and the trend begins again.

Figure 5.3 Example of a regression control chart.

doubled. Checking past scrap reports, they found this same increase every year for July to September. A little thinking made the cause clear; this is vacation time for most people. People with the most experience also had the longest vacations and summer replacement help was unaware of the methods used to contain scrap rates. Therefore, management made certain that summer temporary help was trained in SPC and problem-solving methods before being hired. In addition, each supervisor was told to educate temporary employees on scrap reduction techniques that were now being documented by the regular work teams.

Sudden Changes

Sometimes a chart begins in statistical control, but then the plots suddenly jump outside the control limits. "What makes this jump distinctive is that the pattern of plots could be contained within the control limits. On an X-bar/R chart the averages will leap out of the control limits, but the range chart will remain stable. What has happened is that the

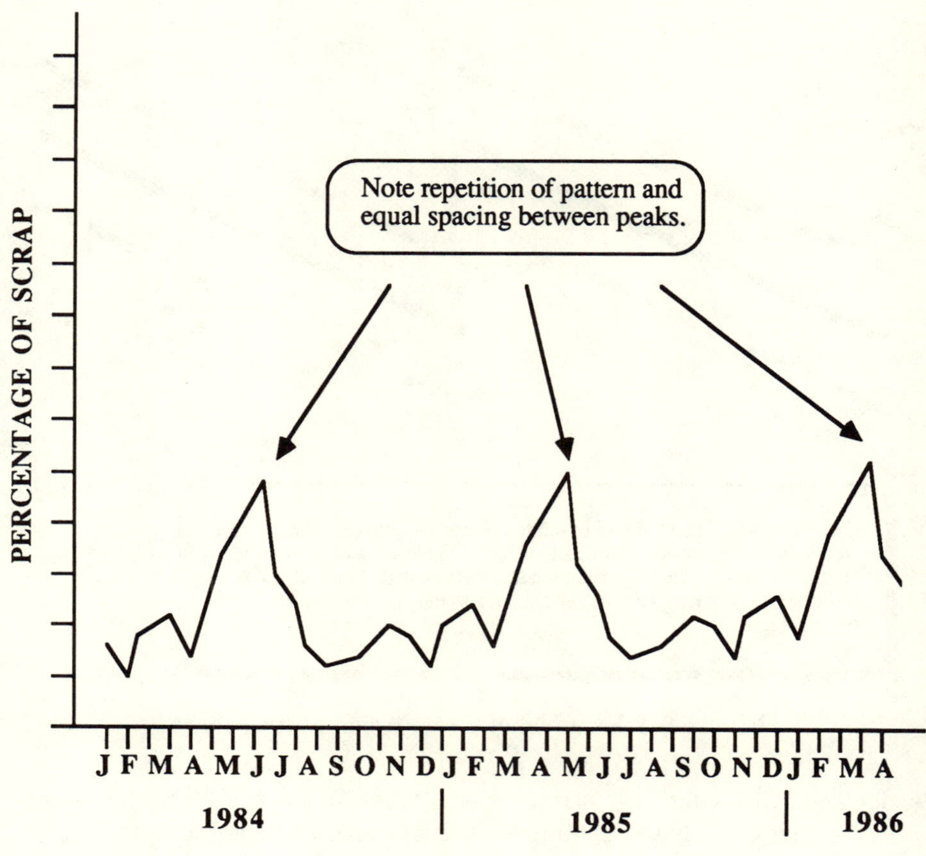

A seasonal trend is a pattern that repeats over a constant period of time, such as days, weeks, months, or years. A seasonal pattern indicates that a constant effect is felt on the process as a function of the season. For example, the above chart of scrap rates indicates that scrap is higher toward the beginning of summer. Possibly because of the high number of vacations during that time.

Figure 5.4 Example of a seasonal trend.

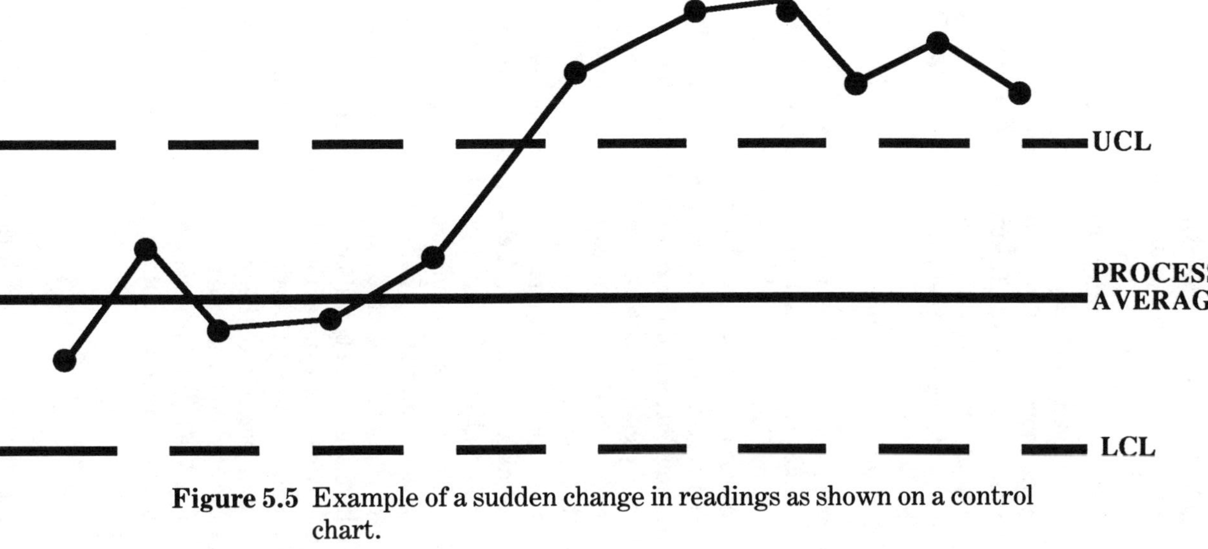

Figure 5.5 Example of a sudden change in readings as shown on a control chart.

process average has suddenly changed. Specifically, something in the process has suddenly changed. For example, the roll of steel feeding a metal stamping machine was changed or the chemical mixture for a cosmetic was altered.

This situation is easy to identify. In the typical case a single, easily found factor has changed and thus, upset the process average. However, corrections may be more difficult to work with. In the metal stamping example, there may not be a way to control the thickness of steel rolls. Instead, a system of adjusting the process may have to be employed to correct the process after each steel roll is mounted for use.

Extreme Variations

In some cases, a control chart was created using the 30 piece process potential study data. Once posted, the plotted points fluctuated wildly across the chart. In other cases, this same series of extreme variations occurs after a control chart is posted for the first time. Both cases are the result of too little information being used to create the control limits. The actual variation pattern of the process was seen only after the chart was posted. The 30 pieces of data underestimated the amount of variation to expect. In such a case, a longer capability study would be required before posting the chart.

In other cases, the control chart may be inadvertently monitoring several processes disguised as one. A classic case is found in a paint shop for small parts. A continuous paint line subjects small metal parts to a coating of paint. The process specifications dictate that each part should have two millionths of an inch of paint over the entire surface. Taking measurements of randomly selected surfaces reveals a pattern of fluctuation. Further investigation shows that the paint line is coating a wide variety of parts. After one type of part is hung on the paint line, another type of part is loaded. The result is that each type of part coats differently under the consistent settings of the paint sprayers. To solve this problem, the paint team creates a chart for each type of part run through the line. Separate readings are taken for each type of part. The results show that each part is capable on its own, but unpredictable when seen as a group of different parts.

No Variations

Sometimes a chart will exhibit little or no variation in its plottings. There are some common causes of this situation. The first is that the tools used to monitor the process are not sensitive enough to detect variations. For example, a scale may weigh to within half a pound, but the process

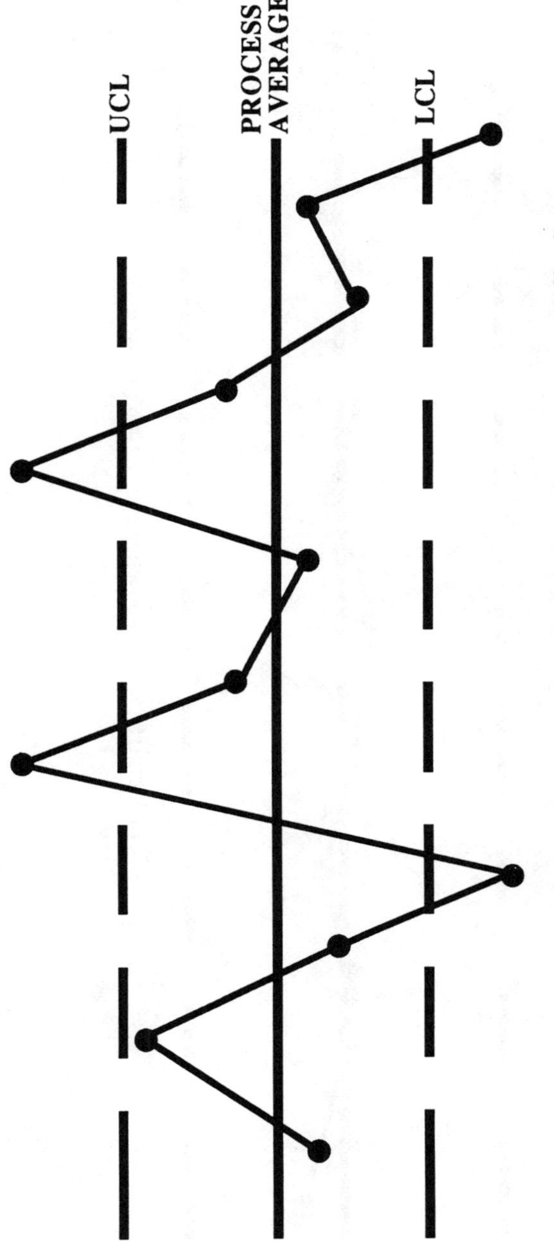

Figure 5.6 An example of a process with extreme variation, difficult to find any statistical control in its pattern.

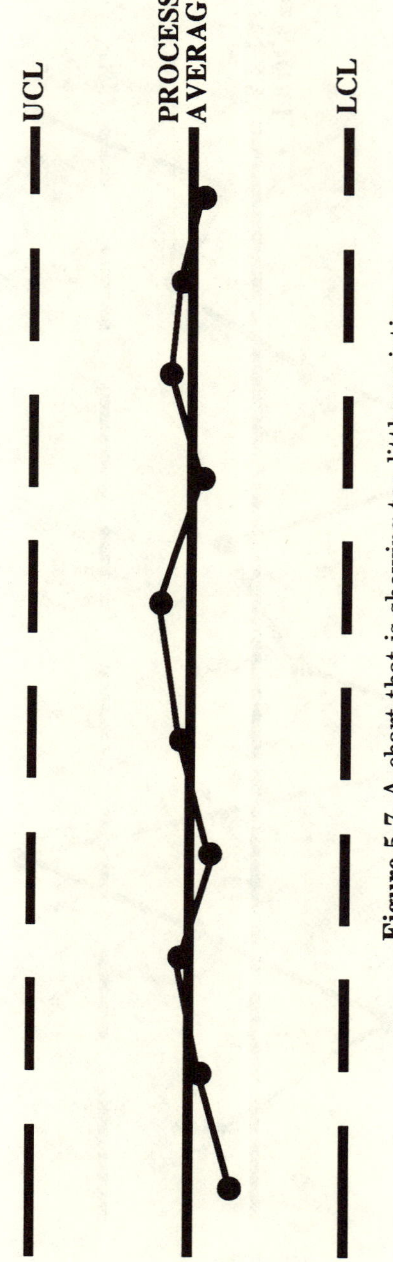

Figure 5.7 A chart that is showing too little variation.

varies on a scale of hundredths of a gram. Evidence of this problem on an X-bar/R chart is seen when the ranges are a mixture of zeros and ones. This also collapses the control limits for the averages. In this case, a new chart has to be made using more sensitive measurements.

Another possible cause is a mistake made by the operators in filling out the chart. When not properly trained on coordinate plotting, they may just plot points where they think they look best. Although possible, it is not the most likely cause of the problem. A stronger possibility is a supervisor who wants good-looking charts. Such a supervisor may be telling the operators to plot only when they find a sample with an average near the process mean.

One rare possibility is that the process has improved dramatically. Although unlikely, this possibility should never be ruled out until a confirming audit is performed, which is done by randomly selecting larger samples of the production and testing to see if the standard deviations are now significantly lower (see Chapter 6 for detailed audit instructions). Should the process actually have improved, then new control limits will be required after the work team identifies the cause of the improvement.

Visible Patterns

Any time that a recognizable pattern is seen on a chart, suspicion should be cast on the statistical control of the affected process. The reason is that randomness usually makes no identifiable pattern for human beings. One example comes from the welding department of an automobile manufacturer. The X-bar/R chart for the overall length of a welded frame created a saw-tooth pattern.

The supervisor investigated the situation with startling results. Each hour two operators would rotate doing this job. The fixture holding the parts together was a little loose, so the parts were not securely held. The first operator was left-handed and pushed the parts together to the right. The second operator was right-handed and did the opposite. The results were two consistent, but different sizes for the finished frame. The supervisor corrected the problem by having the fixture repaired.

Incapable Processes (The Three Choices)

Sometimes, a chart will be successful upon posting. The plotted points will prove to be both stable and predictable. Unfortunately, it will then be discovered that the process is not capable of meeting specifications. This situation requires that three progressive countermeasures be tried.

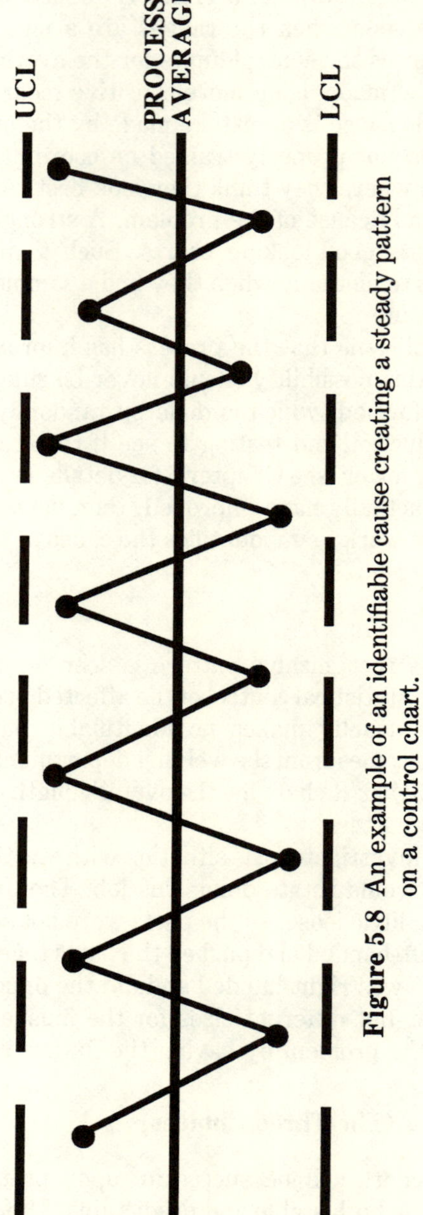

Figure 5.8 An example of an identifiable cause creating a steady pattern on a control chart.

1. Change the Process. Every effort is made to decrease the variations created by the process so that the difference between parts stays within the critical specifications. Problem-solving efforts, new equipment and preventive maintenance are possible sources of improvement. However, a set time limit is usually in place for improving the process. If it cannot be improved, then the next step is taken.

2. Change the Specification. Negotiations with customers or the production planners may uncover that the specification was set without knowledge of existing capabilities. For example, the engineers designed a hole to be drilled with a tolerance smaller than the capabilities of the best machines available worldwide. Such obvious mistakes can be quickly corrected with engineering updates. However, if the tolerance is critical to the success of the final assembly and the process can not be improved, then only one decision is left.

3. Make or Buy? Management must now decide whether it will be economical to proceed with manufacturing this part. Will the potential profits be eaten up by costly rework or sorting expenses? Would it be cheaper to hire an outside firm with better capability to do this part? These types of decisions can be easily made using the data from accounting and the SPC system. The rate of scrap can be estimated from the process potential study.

Overly Capable Processes

There actually is no problem with a process that far exceeds expected levels of quality. In fact, this is the objective of a complete SPC system. However, such processes can create a secondary problem. The illustration of a highly capable process will show the cause of the problem.

What happens is that when statistical control is lost, the parts are all well within tolerance. The past psychology of manufacturing comes into play. The operator does not see why a process that is still producing high quality parts is in need of adjustment and completing a reaction plan. The reason for the confusion is a tendency to focus on the present and not the future. If the first problem goes unchallenged then, others will follow and add to the variations. Eventually, the process will produce parts out of specification. Instead of one problem to address, several are knotted together making the situation difficult and costly to correct. Furthermore, maintaining the high capability maximizes the competitive strength of the process. Such a process is robust against unexpected

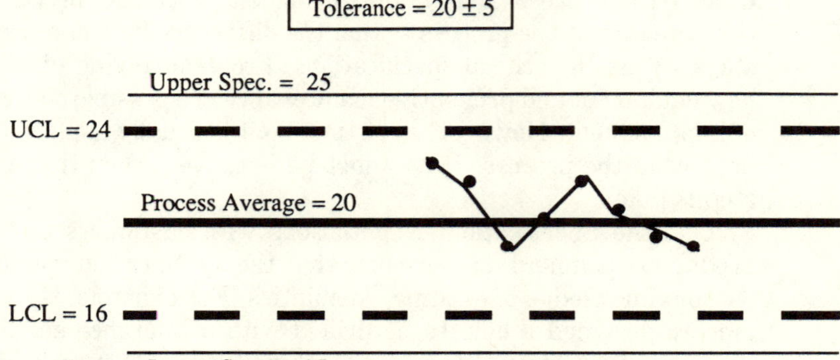

Although the above chart is within specification limits, the process is not capable. The process has an average range of 6.93. Dividing this by the d_2 factor yields an estimated standard deviation for the process of 2.98.

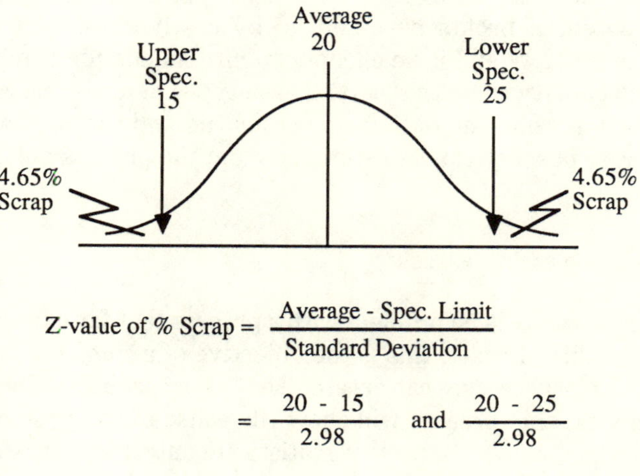

$$\text{Z-value of \% Scrap} = \frac{\text{Average - Spec. Limit}}{\text{Standard Deviation}}$$

$$= \frac{20 - 15}{2.98} \quad \text{and} \quad \frac{20 - 25}{2.98}$$

$$= 1.68 \text{ and } -1.68 \text{ (both equal to 4.65\% scrap)}$$

Therefore, the above control chart is in statistical control, but produces a total of about 9.3% scrap, even though the control limits are inside of specification limits.

Figure 5.9 A stable and predictable process that is not capable.

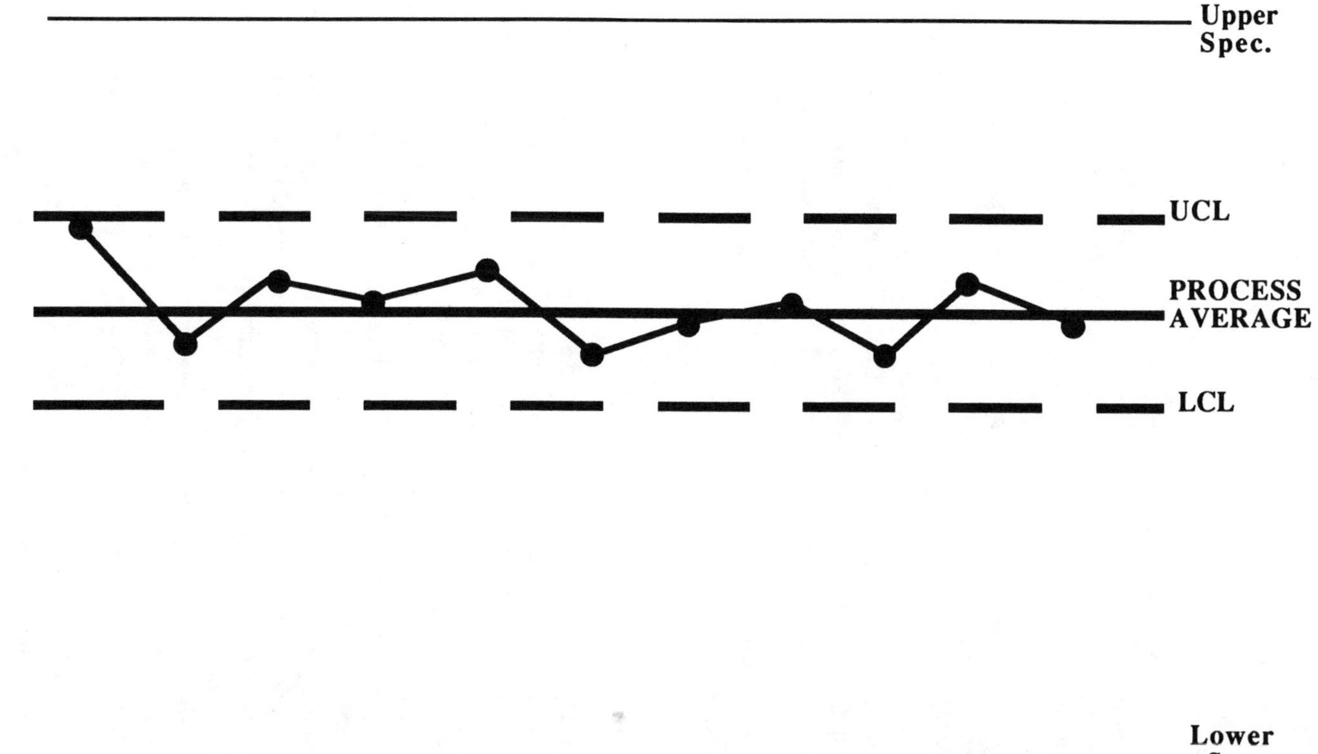

Figure 5.10 A process that exceeds capability requirements.

changes in the environment or materials and actually prevents some problems from occurring.

No Control Possible

Once in a while, a process will be found where the operators have no real control. Adjustments seem to have no effect on the desired critical characteristic. In the worst case, the process cannot be adjusted. The critical characteristic is a matter of luck. For example, a customer of a foundry may request a certain percentage of carbon in the iron. Unfortunately, once smelted, there is very little range in the adjustments that are possible for carbon content.

For such situations, the company must either choose a critical characteristic that can be controlled or study the process in greater detail to find another method of control other than SPC. Experimentation with the method of production could reveal which factors affect the process and the critical characteristic. Perhaps, the foundry people discover that consistent sand moistures in their casting molds will stabilize the amount of carbon in the finished iron. An automated water flow control with sensor is then installed to keep the molding sand to a specific percent of moisture.

Non-normal Distributions

In Chapter 2, one step of the process potential study was to see if the raw data from a capability study form a normal distribution. Only when a distribution of the raw data is normal can conventional statistics be used to estimate variation and later create the control limits. The other reason for checking is to look for hidden causes of variation that can be found by examining the distribution of the raw data obtained on the first posting of a control chart. This data can be quickly analyzed with a histogram or capability charting paper.

A bimodal distribution is created when two processes, techniques, machines, operators or materials are combined in one stage of production. For example, a twin cavity mold of the same part may create a bimodal distribution of a critical dimension. The most likely cause is that the two cavities produce two slightly different versions of the same part. Repairs based on the data should correct the problem. The double bumps of the distribution should then combine to form a more capable normal distribution of production parts.

The rectangular distribution is a clue as to how parts were produced. If the distribution of data for a critical dimension from a supplier is found to be rectangular, chances are the product was sorted and reworked. The

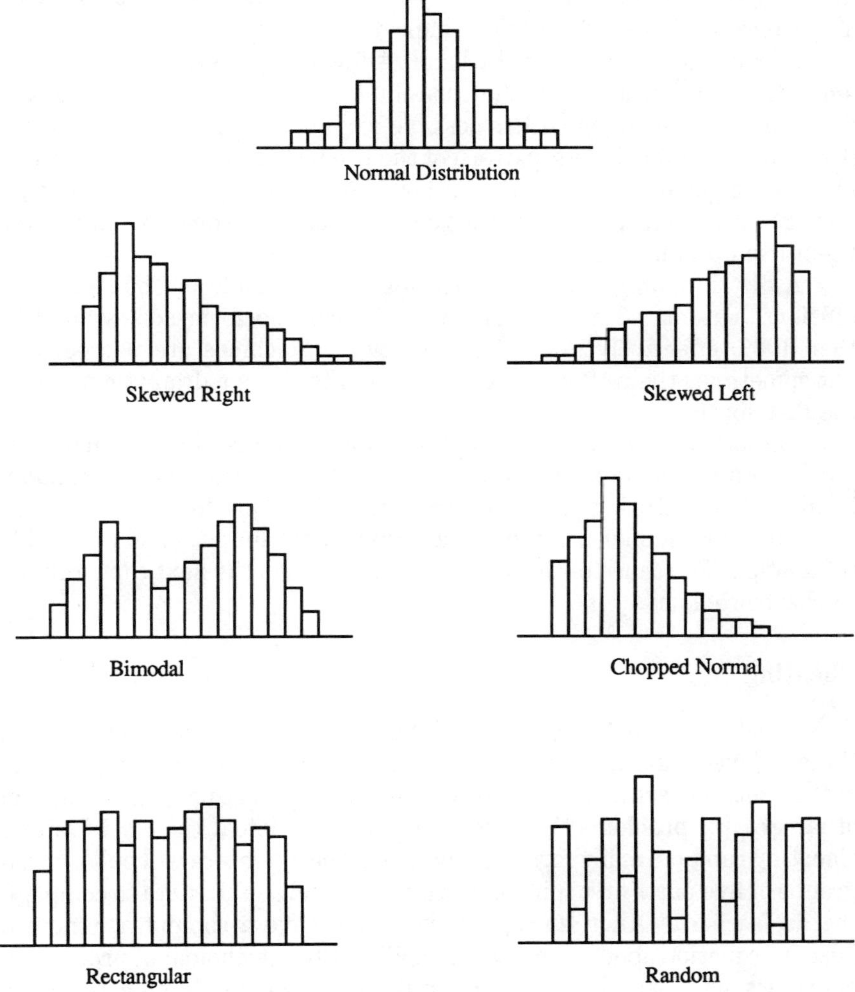

Figure 5.11 Some possible distributions found within a process capability study.

out-of-specification parts were altered to just conform to specification thus, flattening out a normal distribution into a rectangular one. Encouraging the elimination of the need for rework by reducing the sources of variation is the solution to this situation.

A skewed distribution can be created by several possible factors affecting the process. One possibility is that the skewness is away from a specification limit, which might indicate that out-of-specification parts were sorted from the process by hand. A few pieces were missed and the small tail thus created. Another possibility is that machine setting was frequently changed. One other possible explanation is that skewness reflects the natural behavior pattern of the product. For example, a type of electrical component can survive at low voltage, but is destroyed by high voltage. Thus, exposure to higher voltage by some of the parts eliminates them from production.

A distribution with one tail missing entirely could be the sign of an efficient, automated sorting process. Humans perform somewhat less than 100% effectively in sorting operations. Machines are not perfect, but sometimes their effectiveness is enough to make a dramatic mark on the distribution.

One other interesting distribution is random distribution that does not have a definable shape. Each bar on the histogram is of a random height. Manufacturing processes are incapable of producing such a pattern. Just like humans, machines are dependent on patterns to be able to function. The cause of this type of distribution is the next topic related to chart problems.

Cheating

Cheating on control charts is all but unheard of in companies where the employees and management have a clear understanding of the role of the control chart in continuous improvement. Cheating is the symptom of far greater problems than someone trying to defeat the SPC system. Cheating can take many forms. Sometimes, the operators will pull a dozen groups of five parts first thing in the morning and use the data to complete the day's chart. Other times, suppliers will have a computer generate false information about a process capability. Stories abound of production managers who salt defective materials into good production to evade detection under lot sampling plans.

The unfortunate thing about cheating is that those responsible are likely to be caught by a complete SPC system. The test of capability is verified by the audit of later production runs. The charts are monitored by supervisors and audited by management. Incoming materials are subjected to statistical analysis, such as the test of distribution pattern, that will quickly locate problems. The true solution to cheating is to treat the incident as the symptom of a greater problem. Why do the employees feel uncomfortable with SPC? Why can't operators sense the advantage of using control charts? What has management done to encourage suppliers to falsify information? In short, management of the company must address these questions and respond aggressively.

Wrong Chart

The final problem is a simple, but common one—the wrong chart was used for the process. A company may start by examining the defects on a production process only to find out that a critical dimension is the largest defect and needs an X-bar/R chart. At other times a company may attempt to use an X-bar/R chart on the pH of a chemical every hour, when a chart for individuals would be more appropriate. And there is always the possibility that a control chart is not the right choice for a process. A chemical reaction might have to be controlled by balancing several factors. An electronic spreadsheet on a small computer could be used to calculate these balances based on a real-time data gathering probe in the process.

PART 2: PROBLEMS EXPERIENCED WITH AN SPC SYSTEM

Control charts are usually troublesome only during their introduction to the manufacturing environment. With a little training and practice, the charts are soon filled out efficiently and most problems are solved quickly by support personnel. However, the entire SPC system takes a longer period of time to be adapted to the manufacturing requirements of the company. A list of possible problems would be endless. Instead, the major problems reported by companies that have implemented SPC are examined. The emphasis is on the steps that can be taken to prevent these situations.

Some of the major difficulties encountered by companies include:

- Lack of management commitment
- Lack of support from line supervisors
- Inadequate amount of training
- Interdepartmental conflicts
- Failure to gain union support
- Specifications
- Quotas
- Implementation either too fast or too slow
- Documentation
- People
- Supplied material

- Customer demands
- Money
- Quality gurus
- Mistakes
- Good parts
- Data and paperwork blizzard
- Poor quality or remotely located gages

It is rare for a single company to experience all of the above problems. However, it is wise to take steps to prevent each problem to help speed the period of transition to a complete SPC system. There are five basic groups of counter measures. Particular applications will be mentioned below. Usually, using all five will prevent or reduce many of the problems listed above. The five countermeasure groups are training, commitment, planning, communication, and audits.

Lack of Management Commitment.

This problem is almost fatal to an SPC system. For one reason or another top management of the company is not committed to the goals of SPC. Sometimes, this is caused by a disagreement among top managers on how to best implement the system. In other cases, management does not fully understand its role in the system (see Chapter 4).

Lack of Support from the Line Supervisors.

Second only to the lack of management support in its ability to kill a newly created SPC system is the lack of support by the first line supervisors. These supervisors are in the best position to keep the system from being implemented. They are in charge of the operators who must fill out the control charts and document the actions taken. If these supervisors do not believe strongly in the merits of SPC, then their never ending emphasis on production quotas will squeeze out any time for participating in the SPC system. Problem-solving groups will be unable to schedule time away from production to work on problems. In short, the SPC system will stop at the production floor.

Inadequate Amount of Training

To successfully use SPC the participants in the system must be properly trained. For example, an SPC Coordinator should have at least 100 hours of training in statistical methods. However, training is frequently considered an expense instead of an investment in higher efficiency. Thus, many managers limit their training budgets. This is further complicated because a good SPC system requires every employee of a company to receive some SPC training.

Interdepartmental Conflicts

Company politics and the addictive effects of power will eventually infect any company and create divisions among departments. The larger the company, the faster and deeper these lines of division grow. An SPC system cannot survive unless all departments are joined by a common purpose. To achieve this goal, top management begins by agreeing on a common vision in its strategic plan. To spread the idea, all of the benefits of position and power within the company must now reward the advocates of the common vision i.e., profit sharing plans that share the savings from increased productivity, or promotions based on achievements in quality of product and people.

Failure to Gain Union Support

Trade unions are only a problem when they are left out of the planning and implementation phases of SPC. They too will be resentful if not allowed to fully participate in the system. The upgrading of employee skills makes the union membership more competitive in the work place. Many unions now aggressively pursue the education of their members into the new technologies. Several recent union contracts have called for retraining displaced workers and the general upgrading of current skills. General Motors and the United Auto Workers have agreed to a one billion dollar fund for this cause.

Specifications

An obvious problem is when specifications are unrealistic. Some designs include specifications that are impossible to achieve with existing plant equipment. A less obvious problem is the "spec. mentality," that is the idea that if a part is somewhere within a given tolerance, then it is perfect. The idea of making parts "just good enough" has been drilled into manufacturers' minds for decades. Now, SPC systems require people to know more about a product than whether it meets specifications (see Chapters 9 and 10).

Quotas

There is no doubt that modern industry is experiencing a renewed interest in quality control. However, there is no quality control where there is "quantity control". In other words, it is very difficult to produce higher quality when manufacturing procedures are emphasizing maximum production. Piece rates are an excellent example. Few workers will take interest in stopping to fill in a control chart when they are not paid for that time. Another form of this problem is management who uses fixed numeric goals. For example, this year's goal is to increase production 3%. Chances are that any improvements found will be held to no more than 3%. And why not—where is the reward after 3%?

Implementation Either Too Fast or Too Slow

The speed of implementation is difficult to estimate for a particular company. However, the guidelines are well-established. Implementation cannot be so fast that charts appear on every machine at the same time. Charts have to be installed at the rate of the training of operators and their support teams. Otherwise, so much is happening that people quickly lose interest in the flood of activity. On the other hand, implementing too slowly will also destroy a good SPC system. Progress and benefits must be demonstrated on a regular basis. How much and how fast are questions dependent on the type of company involved. Someone working at a large Ford Motor plant is accustomed to seeing projects implemented over the course of years. Yet, a worker at one of Ford's small supply companies will become frustrated when it takes six weeks for a capability study to be completed.

Choosing the appropriate rate of implementation should be directed by management and usually with the assistance of outside consultants. Consultants can be any people familiar with implementing an SPC system, from a professional quality consultant to the president of a similar company. Management must walk the floor every day to get the feeling of the workers' attitudes towards the system. A series of complaints about being left out of the system usually means implementation should be sped up. A wave of complaints about how the production schedule has been destroyed usually indicates implementation should be slowed.

Documentation

It is critical to an SPC system that control chart logs and process logs be kept up-to-date. Many times the system of production distracts the participant from completing needed documentation. This is critical to the success of the entire SPC system. Illustrations throughout this book will point out that the strength of SPC is the ability to trace what happened to improve or downgrade a process. Improvements can then be exploited and downgrading can be corrected.

People

People are the most valuable resource any company can have. However, many times people are not given the tools or training needed to get the job done, causing morale to drop and cooperation to be nonexistent. Participation by management in resolving problems of the production area and providing good training programs will promote development of human resources. One of the true sources of quality is the development of the people who convert skills and materials into finished goods.

Supplied Material

One of the early problems detected by an SPC system is that some of the purchased material from outside vendors is not adequate. The problem is how to encourage the vendors to become more aware of the company's requirements. This is particularly difficult when a vendor is much larger than the purchasing company. Why should a large company pay attention to a customer who represents less than 5% of its business?

Some companies have solved this problem by having supplier awareness days. Suppliers are brought into the company and shown the benefits of SPC. Especially important is the demonstration of the cost savings achieved by the process of continuous improvement. So that lack of knowledge won't cause the supplier to implement SPC incorrectly, the host company offers free training to interested suppliers. Of course, writing new contracts that require SPC can also help to sway suppliers into the SPC camp.

Customer Demands

The implementation of an SPC system should never be rushed, but the realities of the world do not always cooperate. For example, a company may have a major customer decide to require SPC within 90 days. These situations must be avoided where possible and handled delicately when unavoidable. The best method to gain more time is to plan the entire implementation, form committees, hang some charts and then, show the progress to the customer with a request for more time.

Another possible customer demand is to call for a system that is beyond the resources available at the company. If a customer demands that computers be used in all phases of the system for data gathering and the company has neither money nor people to buy and maintain a large number of computers, this situation must be documented and presented to the customer.

Each customer demand that is beyond the reach of the company must be documented. Nothing is gained if a company approaches a customer without information to back up the argument for a change. And, of course, the final option for any company is to calculate whether the demands of the customer can be met or whether they will hurt the company.

Money

Specifically, the lack of money can cripple SPC implementation. Personnel, training, tools, time to chart and time to solve problems all cost money. Without a good budget to support the implementation of SPC and to carry the company through the period of transition, the SPC system will harm the company. The solution is to obtain the funds for completing the task. The first source is the government. State Departments of Labor, Commerce, and Education may have funds for hiring and training personnel. Equipment costs are a matter for tax accountants. Another source of funds is the customer. Some customers are now freely sharing resources with supply companies to help them implement new manufacturing technologies.

There are ways to reduce the needs for funding as well. One method is to tie the SPC implementation to the development of other technologies. For example, a Just-in-Time delivery system could include SPC components within its budget. Another method is for companies to join forces in implementation. A recent example of such cooperation includes one group of companies that hired a single SPC instructor to provide training at a single location. Each class had people from all participating companies which reduced expenses by maximizing the use of the instructor. From this collaboration grew the free exchange of information about implementation problems. One company's solution was shared with the other participating organizations.

Quality Experts

A recent phenomenon is the appearance of well-noted experts in the field of quality. They offer to help a company implement an SPC or other similar system for a charge ranging up to $10,000 a day. Their advice and help is very good, but for smaller companies they can quickly drain resources. Unless the experts' appearance at the company is going to be used as part of a competitive strategy, then the high priced consultants should be avoided. SPC is not magic; any competent consultant can help any company, and competency is demonstrated through success stories and recommendations from other companies.

Mistakes

People make mistakes and there is no way around this fact which means that control charts will have some errors on them and that documentation and related paperwork will not always be complete. The audit procedures described in the next chapter will detail how the sources of such mistakes can be found and eliminated.

Good Parts

If a process produces parts that always satisfy the customer and creates no scrap, why should a control chart be required? This is the type of question that management must be ready to answer. The response is that there is no telling when this leniency will end. Once this form of a free ride ends, the company will have a long road of improvement to meet the new requirements. A company that is striving for continuous improvement will keep correcting its process before customer requirements change so they will be likely to retain the contract.

Data and Paperwork Blizzard

An SPC system could easily lead to extensive paperwork and a mountain of data. The accumulation of control charts, logs, studies, blueprints, and instructions on the shop floor could become overwhelming. The secret of avoiding this problem is organization. No work station should ever have more than one SPC package at a time and other SPC packages should be kept filed according to part number. Charts that are filled in should be sent to quality assurance or engineering for further analysis.

Another part of this problem can be the sudden appearance of statistics in every discussion. Some people with a good working knowledge of statistics may choose to intimidate others by using terms such as operational curves, regressions, beta errors, and phi. The company can avoid antagonism by instructing everyone in statistics. Specifically, statistics can be correctly stated in plain English rather than in mathematical terms. The purpose of statistics is to convert unintelligible quantities of information into meaningful summaries of data, in plain English. People should be encouraged to attack those who fail to use the "plain English" rule. For example, a person does not need to say, "The X-double bar is four with a three sigma of 1.5." Instead, a person should say, "This machine cuts an average part four inches long and I estimate that the variation possible while cutting could be as much one and one-half inches longer or shorter."

Poor or Remote Gages

Repeatability studies will find gages that do not accurately measure parts. Also, some types of measurement require the part to be moved to a remote location such as for destructive testing of a product. Gages that fail to accurately repeat a measurement must be either repaired or replaced. Chapter 9 will examine the methods used to find such gages.

Gages that create delays in measurements affect the efficiency of the SPC system. The closer to real time that data is collected, the more value the information will have. Automated gages or the use of faster measuring gages alleviate this kind of problem. However, some processes cannot be measured quickly, and dependency on control must switch from the chart to the design of the process. Chapter 10 details this procedure more completely.

PREVENTION OF SPC RELATED PROBLEMS

As mentioned before, five basic actions—training, commitment, planning, communications, and audits—can help prevent the above situations from occurring. Management is responsible for the successful implementation of an SPC system. No magic or elaborate management techniques are involved and specific applications were mentioned above. The following is a discussion of these five methods in general:

Training

A well-coordinated training program is the best preventive measure a company can take. A wealth of literature on how to coordinate and conduct successful training programs is available. However, an SPC program has some unique aspects that are discussed below and should be fully considered before training begins.

An SPC system is unique because it requires all employees to be trained and almost all of them to apply their newly acquired skills immediately, meaning that different groups of employees will have different interests in the training material. For example, the engineering group will want a more technical overview than the hourly workers.

Another point is that SPC training really represents a retraining of a work force, that is, the current methods of operation are being replaced by a newer method. As with any change, resistance can be expected. To reduce resistance, a company needs to actively distribute information about the planned training program. Management could explain that support groups and tutors would be available to all training participants to ease the anxieties of transition (see Chapter 15).

Knowledge does not end with training and most of the learning that people experience with SPC occurs after the training. It happens when employees observe the control charts first hand on the production floor. Resource people have to be available to answer questions and sometimes, the participants demand further training in related areas to help expand the benefits of SPC. Topics such as math skills, blueprint reading, or geometric tolerancing are some examples.

The difficulties of training are part of the price for the successful implementation of SPC. A good training program will be tailored to the needs of each participating group which will emphasize each member's role in the SPC system, give each individual practice at participating, and instruct each person as to which people to go to with questions.

Commitment

The first step in implementing any SPC system is to obtain and keep the full commitment of upper management. With upper management commitment, the traditional roadblocks to SPC can be more easily removed. In addition, planning and resource allocation can be smoothly coordinated.

Commitment to SPC is obtained when upper management is convinced that SPC will greatly benefit the company. When a major customer is going to require SPC on future contracts, the benefits of having a system become obvious. Under other circumstances, the job of convincing management can be more difficult. Extensive economic reports about the benefits of SPC must be circulated among management frequently. Evidence of the usefulness of SPC must be demonstrated and documented. Some companies do this by testing SPC methods on a single problem job. Whichever method is used, the bottom line is the main consideration for management.

Planning

Good planning will prevent conflicts and shortages of resources. A good SPC implementation plan will include the roles that each manager, supervisor, and employee will play in the system. In addition, the implementation plan will describe the goals, objectives, and timelines. The resources needed to achieve these items will also be noted. Such a plan should be widely published within the company. Revisions should be allowed only through a set series of procedures so that the authority of the plan is established and maintained.

Key points within an SPC implementation plan include the role of first line supervisors, engineers, and operators. SPC systems will fail when the first line supervisors do not have a clearly defined participative role in the system. Management actions must also reinforce the importance of the first line supervisors. These supervisors are the ones who will be enforcing the use of the SPC system. They will also be responsible for reporting where added resources are needed. The role of the operator is to follow the first line supervisor's lead. The operator must fill out the charts and notify the supervisor when statistical control is lost.

Communications

All communications about SPC must be plain and simple. Extra wordage will only lead to confusion over the purpose and use of SPC. If the strategic plans call for SPC to meet customer requirements, then top management should continually discuss with employees the requirements and how the company plans to meet them.

Another level of communication that must be present is among work groups. A problem-solving team must have the ability to communicate freely with other work teams or management. Such communications will promote the smooth implementation of solutions in each department.

Audits

Auditing confirms the accuracy of the SPC system and finds new potential for improvement. A regular system of audits is required. This is not an IRS type of audit. The idea is not to find out who's not performing well but to explore possible opportunities. An auditor may find that most people in the plant do not correctly calculate a range, so, management sets up additional training, using tutoring. This action encourages development instead of punishment (see Chapter 6).

A regular audit of the gages will find either those inappropriate for the job, or those which do not measure accurately. The audits will also detect differences in the methods people use to measure, helping to establish measuring standards to prevent errors in data reproducibility.

Audits of the charts, documents and gages will also serve as a continuing reminder to the employees of the company's commitment to quality and SPC (see Chapter 9).

SUMMARY

Every system will have its share of problems, which usually cannot be foreseen. The period of transition to SPC is marked by a flurry of problems that range from the critical to the insignificant. Strong commitment, planning, training, auditing, and good communication will prevent many of them. The remainder are solved by a willing and flexible management. After the transition period, the company will have the opportunity to expand the SPC system more deeply into each department. Therefore, management must deal with each unexpected problem as it occurs. The solution is always available and the best one benefits everyone involved. Then, instead of just solving a problem, an opportunity may be found. Control charts demonstrate this principle. Each of the problems described above about control charts could also be seen as an opportunity.

SPC does not come custom fitted to a company. Instead the company will unearth a series of difficulties during implementation. By eliminating these previously unseen problems, the company becomes more competitive. The alternative is to have a downturn in the economy reveal the same problems when it is too late for the company to react.

CHAPTER 6
AUDITING AN SPC SYSTEM

Control charts should be checked frequently for accuracy and completeness. The entire SPC system should receive a detailed formal audit at least once a year. Management should assign trained auditors to this task who will check not only accuracy, but the conformance of the current SPC system to the objectives stated in the quality plan. A separate group of auditors will examine the SPC systems in place at the supply companies. The purpose of the audit will be to make sure that purchasing contract requirements for SPC are being met. In addition, the auditors will be part of a process to encourage the supply companies to create SPC systems that can easily coexist with the in-house SPC system.

Reason for Audits

There is only one reason for auditing—to create opportunities. This idea may sound strange to people who are used to auditors being part of a regulatory body. However, to make auditors effective, they must never have the ability to carry out disciplinary actions; that is not their function. The purpose of an auditor is to accurately portray the present condition of a business or part of that business. This chapter will focus on the role of the auditor as a management function. Specifically, how to audit the Statistical Process Control (SPC) system.

After an SPC system is in place and training has been completed, a regular system of auditing should be established. Many people do not trust an SPC system which is a good instinct, since many systems have serious flaws when they are new and people are inexperienced.

PART 1: AUDITING THE IN-HOUSE SPC SYSTEM

Where to Start When Auditing. Management must begin by listing the members and stating the intentions of the audit group. A typical audit group is composed of an audit review team who oversees audit operations, and the actual auditors. The auditors will use a select group of tools to accomplish their tasks that are set forth in the guidelines produced by the audit review team.

The Audit Review Team. The membership of the audit review team usually consists of one member from senior management, the quality assurance manager, the production manager, an engineer, and at least one hourly employee from the shop, which provides a balanced team. The team can then name the actual auditors and meet with them regularly. The function of the audit review team will be to plan audit operations and review reports submitted by auditors.

The audit review team begins by drawing up the schedule of places and systems to be audited. This schedule is then distributed company wide so that each affected department can confirm its participation. Once the auditees have confirmed in writing their agreement with the schedule, the assignment of tasks takes place.

The Auditors. The first task for the audit review team is to name the actual auditors who can be professionals from inside or outside of the company. In either case, they must be trained and experienced in auditing methods. If a company wants to audit the cost impact of the current SPC system, then a cost accountant would be one possible choice for auditor. If this same cost accountant is also experienced with SPC techniques, all the better.

Intentions and guidelines for auditors. Typically, the mission of an SPC audit group is to confirm or correct the intended function of the SPC system. It is a good idea to note that cost reduction should be an objective of an audit. Audit members should encourage cost saving measures they find, as well as discourage improper practices. The way to discourage improper practices is to report them to management who must take responsibility for such actions. Hewlett-Packard provides a model to follow and publishes "best practices" instead of a rule book. The book is a collection of "best ways" to accomplish a task. When someone finds a better technique, it is published and distributed. This approach encourages improvement, instead of punishment for poor performance.

Tools of auditing. An auditor has a whole battery of tools available to assist in the audit. For auditing an SPC system, several of these tools are required. The first is the audit checklist. Some possible checklists are shown below. The purpose of the checklist is to help guide the auditor in examining the SPC system. By requiring a fixed list of items to check, the possible bias of the auditor is reduced. The checklist also helps to keep interviews consistent preventing interviewees from feeling as though they were treated unfairly.

Statistics can be helpful for the auditor, and statistical summaries of capabilities, capacities, and defects experienced will help the auditor present an easily understandable picture of the process to management. Another way to present a picture of the process is to create a flow chart of the process being audited. Once the control charts are in place, their reported capability and a list of the problems encountered could be noted at each stage of production, helping to locate bottlenecks or other costly problems.

Every auditor must be equipped with the standards that apply to the SPC system. For example, a Ford Motor Company supply company would have copies of Q-101 available for the auditors (see Chapter 14 for a complete description of Q-101). It is vital that the auditor ensure the customer requirements are being met. In addition, the internal procedures for operating the SPC system must be examined against the actual practices in the production area. The auditor must be able to document that the SPC system is being used as intended.

Other tools recommended for the auditor include "human skills" and a sense of cause and effect. An auditor's ability to understand human behavior has twin benefits. First, the auditor makes the auditee feel more at ease. Second, the auditor can observe behaviors that are hard to document, but might be destructive to the SPC system. In other words, the auditor will be better able to detect "attitude problems." Observation of cause and effect will help to clarify the auditor's final report. An incapable process is an obvious problem, but its effect on the next stage of production is not always as clear.

How to Examine the Total SPC System

The method for examining the entire system appears simple, but it requires meticulous attention to detail for success. First, the company must specify the goals of the SPC system so that the audit review team can prepare a schedule of activities for the auditors. People affected by the schedule are notified in writing. Once written confirmation of the schedule is received, the audit can begin.

First Stage of any Audit: The System of Management

The audit begins with the company's strategic plan and its chief executive officer (CEO). The auditor must be certain that continuous improvement is firmly ingrained into the operation of the company. Therefore, the auditor begins by looking for a quality plan that includes goals, objectives, and timelines. One of the goals must be the establishment and use of an SPC system. The SPC system must be part of the production process and not just a task for Quality Assurance. The auditor then confirms the plan is being followed by looking for documentation which shows that the goals of the quality plan are being actively pursued. Documentation might include minutes of management meetings, departmental budgets, or procedures created by an SPC committee.

Next, the managers of each department must be interviewed. Each manager must be able to recite the operating philosophy of the company and his role in the strategic plan. The auditors might also quiz managers on their role in maintaining product quality and the company's competitive position. Later, the consistency of answers among managers will indicate how well management has communicated the needs of the quality plan.

Then, randomly selected operators from every shift and department must be interviewed and asked a few questions. The interview should be short. The key questions are:

1. What is the purpose of a control chart?
2. What is a control limit?
3. What is the capability of the process you are now working with?
4. When out of statistical control, what do you do?

These are not the only possible questions, but they do cut to the heart of the purpose of SPC. Again, answers to these questions will be compared among operators to assess the consistency of the SPC system and how well it has been implemented.

Second Stage: The Great Paper Chain

Inherent in any manufacturing system is paper work. A proper audit will examine the SPC related paper work and the auditor's primary purpose is to confirm that complete records are being kept. In a manufacturing plant an auditor would select several products, and while checking their quality, the auditor would also trace the associated paper work. What materials went into making this product? Where was this lot shipped to? Who made the parts at 9:00 on May 27? If any of this information is missing, then why?

Naturally, what the auditor is looking for in the paper work will depend on the priorities of the company. For example, General Motors supply companies are under great pressure to have flawless lot traceability. Ford Motor Company suppliers need records to support a Just-in-Time delivery system. Defense contractors are under pressure to have cost and reliability information available. Therefore, the auditor should examine paper work systems that meet customer needs first, and the needs of the internal operation second.

Third Stage: The System in General

A Check should be made on the supporting parts of the SPC system.

- All gages should be calibrated by a nationally certified laboratory each year and have a calibration sticker.
- An SPC Coordinator should be present on each shift.
- All employees should have received at least basic training in SPC.
- Inspection methods and test procedures should be written, on file and used with each required inspection.
- Problem-solving groups of some sort should be present, with a channel open to top management.
- Blueprints should be controlled and accounted for.
- SPC information should be used by departments such as Purchasing, Sales, Engineering, Production Planning and Personnel.

Examining the Control Charts. The auditing of control charts should be done quickly every week by the supervisors. This brief audit and a more detailed yearly audit cover the same four points:

1. Documentation. Is all of the information about the chart correctly noted? This can be determined by asking the question, "If I found this chart on the floor one day, could I correctly identify where it belonged and who had been filling it out?" Also important, is the examination of any notes made on a chart. If a stranger read these notes, would they be understandable? If a control chart had one point circled on the front and a reaction note on the back and the reason for the process variation was stated as "technical stuff," clearly, this notation is not the type of documentation that makes a control chart successful.
2. The Picture. In other words, the patterns formed by the plotted points. Are there trends, nonrandom patterns, or zone analysis problems? Zone analysis is a good tool for detecting most of these problems. If a company has a statistician and a computer handy, a finer check can be made using the run-test for randomness, auto-correlation, or moving average analysis.
3. The Data. An X-bar/R chart is an example. Are the averages correctly calculated? What about the ranges? Are points plotted where they should be? These are the three most common mistakes made by the users of control charts. An audit should randomly select time periods on a chart and examine the accuracy of calculations and plotting.
4. The Distribution of Data—A histogram created from the indi-

vidual readings on a control chart can test for a normal distribution. As explained in Chapter 5, non-normal distributions are the sign of problems in the manufacturing process. Also, without the normal distribution, the calculations of standard deviation and capability cannot be conducted.

The Formal Audit. The annual audit of the SPC system must go deeper than the once a week overview conducted by the supervisors. A checklist should be produced to look for the following:

- Is a reaction plan attached to the chart?
- Are points out of statistical control circled, signed by the supervisor, noted on the back, and taken care of?
- Are problems and their solutions well documented?
- Are control limit values and specifications noted?
- Is the Cp or Cpk index on the chart?
- Is every sample noted by date and time?
- Are samples being taken at the rate noted on the chart?
- Are the operation, material, tools, and other parts of the process in used noted?
- Is the last completed control chart nearby?
- Are the blueprints for the part also nearby?
- Are there greasemarks, fingerprints, coffee stains and other signs of regular use on the chart?

This last point is particularly important showing that the operators are indeed, using the SPC system.

Comparing Charts to Parts. It is possible to have a control chart that indicates statistical control at the same time the process produces parts that are of quality lower than predicted by the chart. In the words of one company's president, "I have good charts and bad parts." There are many possible causes for this situation which could be a result of paying too much attention to charting and not enough to the operating characteristics of the process. The auditor must confirm whether the charts accurately portray the process.

A random sample of about 30 parts is drawn from production. The average and standard deviations are then compared statistically with those predicted by the control chart. A simple t-test with either an F-ratio or Chi-square test is sufficient. The t-test is a test for significant difference between two means. The F-ratio and the Chi-square can both be used as tests of a significant difference between two standard deviations. These methods are easily obtained from most introductory statis-

Comparing Two Averages:
The t-test

Any two averages can be compared regardless of sample size using the formula:

$$t = \frac{\overline{X}_1 - \overline{X}_2}{\sqrt{\dfrac{s_1^2}{n_1} + \dfrac{s_2^2}{n_2}}}$$

If the two standard deviations differ significantly, they have to be pooled using the formula:

$$s^2 = \frac{\Sigma(x - \overline{x})_1^2 + \Sigma(x - \overline{x})_2^2}{n_1 + n_2 - 2}$$

Example of t-test:

An SPC auditor samples 30 parts from two machines that are suppose to form a part five inches long. The first machine has an average of 4.99 and a standard deviation of 0.10. The second, an average of 5.05 and a standard deviation of 0.09.

$$\frac{4.99 - 5.05}{\sqrt{\dfrac{0.10^2}{30} + \dfrac{0.09^2}{30}}}$$

or, $\dfrac{-0.06}{\sqrt{0.0006033}}$

thus, $- 2.44$

Significant t-value $= 2.042$ $(.05)^*$

* thus, no significant difference

where, $\overline{X}_1 =$ the average of the first sample

$\overline{X}_2 =$ the averge of the second sample

$s_1 =$ the standard deviation of the first sample

$s_2 =$ the standard deviation of the second sample

$n_1 =$ the size of the first sample

$n_2 =$ the size of the second sample

and, degrees of freedom $= n_1 + n_2 - 2$

Figure 6.1 The t-test for comparing two sample averages.

tical textbooks. Confidence levels should be at 95% (see tables in back of book). These statistical tests help confirm that the chart accurately reflects the variations within the process. Similar tests can be performed for attribute data.

Every variable chart in a plant should have a Cpk ratio calculated. When a process is audited, the Cpk ratio on the current chart should be compared to the ratio on previous charts. The higher the Cpk, the better. Attention should be paid to variation and how well centered the process is within tolerances. For attribute charts, the auditor is looking for down-

The F - Ratio

The F - Ratio is used to test for a significant difference between two sample standard deviations. For example, suppose a control chart currently being kept on a grinding process indicates that the standard deviation for the past few days has been 0.005 mm. The supervisor wants to compare this current amount of variation to the standard deviation on a chart from six months ago to see if the process has been improved. On the old chart is a standard deviation of 0.006mm. Each is based on a sample of 21 pieces.

$$F = \frac{s_1^2}{s_2^2}$$

where, s_1 = the standard deviation of the first sample

s_2 = the standard deviation of the second sample

and, degrees of freedom are -

$$DF_1 = n_1 - 1$$

$$DF_2 = n_2 - 1$$

Therefore, $F = \dfrac{0.005^2}{0.006^2} = \dfrac{0.000025}{0.000036} = 0.694$

$$DF_1 = 21 - 1 = 20$$

$$DF_2 = 21 - 1 = 20$$

The critical values for the F-Ratio are 2.124, thus the standard deviations are not significantly different.

Figure 6.2 The use of the F-ratio to test for a difference between standard deviations.

Chi - Square

The chi-square is a test statistic used to compare a standard deviation from a sample against a specified amount of variation. For example, suppose a bid requires that each batch of product has a standard deviation of 0.020 inches or less. The first shipment has 30 pieces sampled to discover a standard deviation of 0.025. Is this sample standard deviation significantly greater than the 0.020 inch specification?

$$X^2 = \frac{(n-1)\ s^2}{\sigma^2}$$

where, σ = the established standard for variation

s = the sample standard deviation

n = the sample size

and, degrees of freedom = n - 1

$$= \frac{(30-1)\ 0.025^2}{0.020^2}$$

$$= \frac{29\ \times\ 0.000625}{0.0004}$$

$$= 45.3125$$

The null hypothesis would be that the sample standard deviation is statistically equal to 0.020. Looking up the critical Chi-value for 95% confidence with a sample size of 30, a value of 42.557 is found. Since, the Chi-value is higher than this, the hypothesis is rejected. Instead, the sample of parts are seen as having too much variation.

Figure 6.3 The use of the Chi-square to evaluate variation.

ward trends meaning improvements. Having statistical control on an attribute chart demonstrates the capability of producing consistent scrap or rework.

The above information can be summarized in numbers and graphs for use by the audit review team. A list of the processes examined should then be prepared that would sort out capable from incapable processes. The auditor only collects the data; the auditor should play no part in the use of the data. It is up to the audit review team to decide which actions to take with this information.

Fourth Stage: Measuring the Effectiveness of the SPC System

This part of an audit can also be an entirely separate audit process. The important feature is to agree upon the definition of "success" before auditing takes place. The criteria for successful improvement in the process must be clearly defined. Is success measured in terms of improved capabilities, lower costs of production, or higher quality of product? Essentially, the purpose of the audit is to determine the ecomomic impact of the SPC system on the company. Information should be collected on the cost for sorting, rework, defects, and defectives then combined with control charts, so management can accurately evaluate the improvements realized by actions taken on the shop floor. A good cost accountant is necessary for setting up this procedure.

For example, an auditor can report the brackets made last year had to be sorted on an average of every five lots costing $10,000 in inspection and labor for one fiscal year. A $3,000 repair of a stamping press fixed the problem. Clearly, the repair would be reported as cost effective.

A formal audit report of the SPC system could itemize each quality cost into one of the four traditional headings:

1. Prevention Costs—the expense involved in the prevention of problems that could affect the quality of the product or service such as the cost of quality engineering, training, supporting problem-solving teams, and so on.
2. Appraisal Costs—the expense of maintaining stated quality levels including inspection, testing and auditing.
3. Internal Failure Costs—the expense of correcting problems at the source of production. Scrap, rework, and repair are examples of these types of costs.
4. External Failure Costs—the expense of correcting problems after the product has been shipped or the service rendered. Complaints, warranty work, and lawsuits fall in this category.

Final Step in Auditing: Preparation and Presentation of the Written Report

Upon completion of the audit, the auditor or audit team must prepare a final written report for the audit review team, which is a summary of the findings from the SPC system. Both strengths and weaknesses should be pointed out and the main theme should be how close the SPC system is to the goals stated in the quality plan. If a company's main goal is to have an SPC system that meets General Motors' supplier requirements, does the current SPC system meet these requirements?

The final audit report is presented at a special meeting of the audit review team. This meeting is attended by the chief executive officer and possibly other affected managers. The auditor presents the written report to those present, along with a short oral report on the results. It is during this time that the auditor is debriefed. Each member of the audit review team will question the auditor about particulars or will ask for clarification. It is also at this time that nonaudit material may be presented. For example, the market research manager may give a brief report on the SPC systems at competing companies as a reference for the audit report.

Management's Actions. After the report has been presented and the auditor has been debriefed, management must act on the information provided. The first action is to inform the affected departments of the results of the audit. Feedback from the departments is important. After that, the managers must choose whether to modify methods, change parts of the quality plan, or re-allocate resources to help the SPC system towards its intended objectives. These actions must be documented along with the reasons for change so that other managers, employees, or future team members will be able to trace the path of development for the SPC system.

PART 2: AUDITING SUPPLIERS AND NEW SUPPLIER RELATIONS

The relationship between a manufacturing firm and its suppliers is mostly contractual. Therefore, the role of an auditor with suppliers is to confirm that the conditions of relevant contracts are being met. SPC is one of the major concerns of an auditor. Since SPC is a system of management, it must be examined to see that it assures maximum quality at minimal cost. In other words, the auditor is seeking documented evidence that continuous improvement is the operating philosophy of the supply company.

Because of the statistics and legal requirements involved, supplier audits should only be conducted by skilled and experienced professionals. The auditor can be an in-house professional or a hired consultant. In either case, the auditor works best when supported by a preplanned schedule and technique of auditing. The auditing itself is more effective when it takes place before any contracts are awarded. Once a supply company receives payment for its work, its motivation to conform drops. The auditor is also more effective when independent of the company being audited.

Auditors that are assigned to supply companies still have to report to the audit review team. The auditing guidelines and tools are similar to those used by their in-house counterparts. However, the supply company auditors are significantly different in one respect—they have to work directly with the purchasing department. The first written audit report will go to purchasing. The meeting with the audit review team will be for information only. The purchasing department will take any required management actions. These actions will be guided by the purchasing agreements that exist with the auditee.

Step 1: Should Suppliers be Audited?

Auditing is an expensive and time consuming activity and has also been known to create ill feelings among organizations when it is not used properly. The first step for any company considering an audit of its suppliers is to have top management carefully study the true need for auditing. Auditing in itself will not cure quality, delivery, or other problems. Only an aggressive and cooperative supplier relationship will reduce such problems. Therefore, any audit must be seen as part of the complete program of supplier relationships. Auditing is meant for situations where both auditor and auditee expect to benefit from intense examinations of their systems.

One issue to consider is the current incoming inspection methods. If a company decides to require its suppliers to have SPC and will regularly audit the systems, then it must cease to use lot acceptance sampling. Plans like MIL-STD-105D with Acceptable Quality Levels (AQL's) are in direct opposition to the purpose of SPC. What good will it do to advertise an AQL of 1% defective when the company's SPC requirements demand zero defects?

The benefits and intentions of revised audit procedures must be clearly communicated to the suppliers. Information could be presented in the form of a list of goals such as lower inspection costs, fewer defective materials, just-in-time deliveries, or more independence for the supplier.

However stated, the main goal must include improved quality of the production at the supplying and purchasing companies. A good supplier SPC system and outside audits can go a long way towards improving both supplier relationships and performance. Quality at the source helps assure better quality during production. However, a clear message must be communicated about the reason for SPC and auditing. The supplier should be shown that the interest is in finding and solving problems.

Step 2: Re-establishing Supplier Relations

Traditional forms of supplier relationships tended to be based on the idea of regulation. The purchaser was cast into the role of regulating the activities of the supply company. Because these relationships were based on contracts, an adversarial atmosphere was created. SPC thrives on cooperation and trust. To convert the existing supplier relationships to a more open approach, the principle of continuous improvement through cooperation has to be established within purchasing policies.

All new purchasing contracts should require SPC. The contract can state that a 30 or 100-piece capability study be part of the bid. In this way, a company can have a rough estimate of the process potential at each supplier. This requirement also encourages the adoption of SPC methods at the supplier's facilities which is the reason the Cpk index is so important to Ford Motor Company.

A complete survey of capabilities could also be done on-site and would look at the price, process, and delivery capabilities of the selected process. This is a time consuming procedure and one of the reasons that many companies now prefer single sourcing, which makes tracking easier.

By openly sharing information with the suppliers, a company can encourage cooperation. The capabilities are not a trick; they are a source of information that promotes good communication between purchaser and supplier. For example, a company could be told that the new part it is supplying is within specifications, but a critical characteristic has too low a Cpk index. By examining the data together, the two companies may find the process does not center on the optimal specification for this dimension. Comparing the potential improvement and adjusting the process with the capabilities of the mating part enables a company to provide clear information about requirements.

Step 3: One time only Quality Survey

Examining a vendor's quality system only once is a controversial concept. The idea is that a company should trust its suppliers and judge their acceptability by how well their parts work and how well they co-

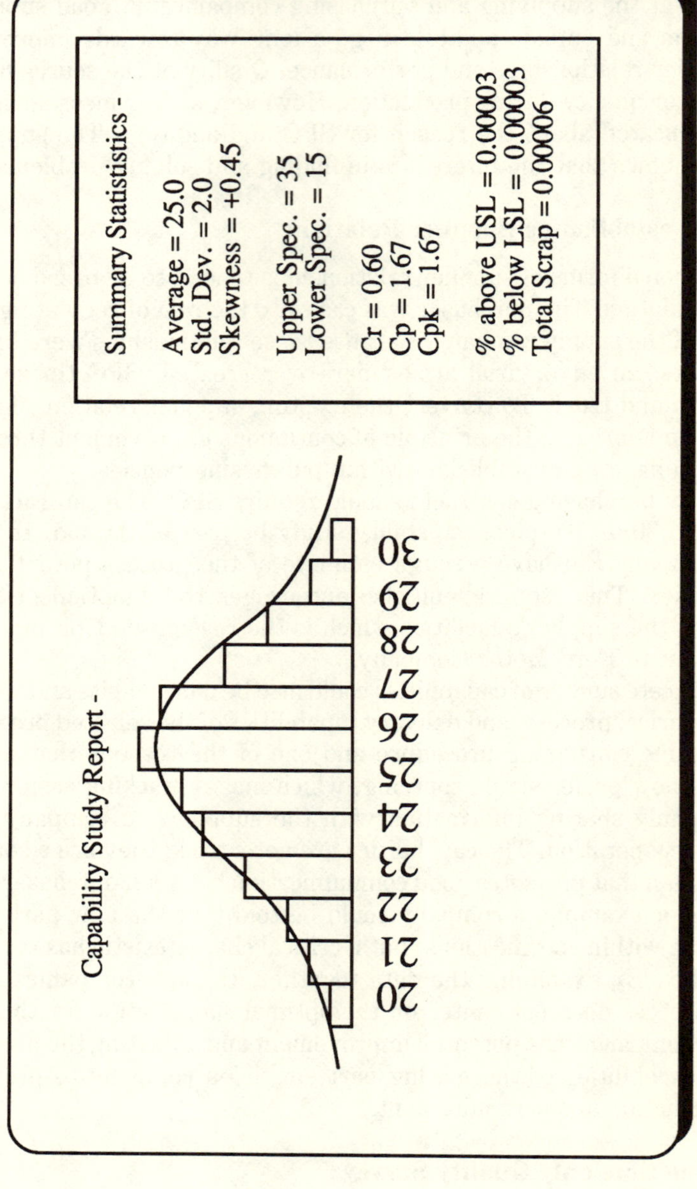

Figure 6.4 An example of a capability study.

operate with the new requirements. The motivation for a one-time-only audit is to save money and personnel by certifying selected suppliers. Repeated and expensive audits are replaced by monitoring the success of supplied materials at the point of manufacturing. This type of supplier relationship relies heavily on trust, cooperation, and communication.

If a vendor is to be examined closely only once, then the focus of the audit will be on the strategic operations of the supply company. The audit begins by examining how the supplier assures quality, involving an audit of the management and operating procedures. The key question is whether continuous improvement is the documented operating philosophy of the audited company. Auditing of the SPC and quality systems is similar to the in-house audit detailed earlier in this chapter.

But, an auditor must also pay close attention to the related areas. For example, are safety and good house cleaning encouraged? Is the company run by good management? Are the employees encouraged to use new methods, are training sessions held on site, and are senior managers visible on the shop floor?

If the supplier relationship is established correctly, there will be no need for a person in the purchasing company to be the one to run to each supplier as problems arise. Instead, the suppliers are brought on-site at the start of each new contract and educated about the critical needs at each stage of production. For example, if threads free of flash are needed for a company to assemble a unit, they should bring the supplier to the shop floor and show why the flash causes problems.

Step 4: Auditing Incoming Material

The goal of many manufacturers today is to eliminate the incoming inspection department. They want suppliers to send in defect-free material with SPC charts that verify the usability of the parts. To accomplish this objective, an occasional audit of the parts is needed to confirm the charts do indeed reflect the quality of the parts, which helps to detect problems, such as differences in measuring between the two companies. Discrepancies occur when the gages at the supplier do not agree with those of the purchaser.

Some managers today think that auditing is for catching suppliers who try to "cheat". Not so. If a company can't trust a supplier, it has a bigger problem than product quality. Besides, why should a supplier, well informed on what's needed, want to produce something else? It's cheaper to make it right.

It is possible to make a Cpk chart for incoming materials. The average and estimated scattering of individual parts can be plotted against the specifications to show both conformance and improvement. A company can even share this information with the supplier. Many supplier problems today exist because the supplier rarely knows what the customer found when auditing the good lots of parts.

As stated earlier, a t-test can be used to compare the average from a sample of incoming parts against the process average on the accompanying control chart. The F-ratio or Chi-square can compare estimated standard deviations.

The removal of a supplier rating system has caused controversy for two reasons. First, any rating system will be biased toward some fixed criterion. For example, most relate the cost of shipping back to defective goods, which weighs against the larger suppliers. The second problem, is that suppliers are encouraged to spend their time working on their ratings instead of continuous improvement.

Step 5: Joint Ventures

The eventual goal of modern supplier relations is to make the suppliers party to joint ventures in production, which means bringing in the best suppliers the moment a project begins. They should help with the bidding for customer contracts. In short, cooperation is a competitive tool. The joint venture has been receiving a lot of good press lately. The reasons are simple. The joint venture encourages two companies to share both resources and information and promotes better cooperation and feelings between them resulting in mutual benefit. Chapter 14 contains a case study of a recent joint venture.

HOW TO ENCOURAGE PARTICIPATION

Only one question remains—how to convince a supplier to participate? Suppliers who are larger than a purchasing company are especially hard to persuade. The answer is persistence. The interested company can offer SPC training to suppliers, have supplier awareness days, and especially, relay SPC success stories. Encouragement and support will be more effective than contract disputes.

SUMMARY

Audits make it possible to confirm or correct the objectives of an SPC System. Audits may also indicate ways to reduce operating ex-

Cpk Control Chart

The Cpk control chart is designed to help companies to monitor various vendors, quickly and efficiently. Upon the receipt of vendor goods, a small sample (usually 30-50 parts) are examined for critical dimensions. The average and standard deviation are calculated.

The average is noted as a circle.

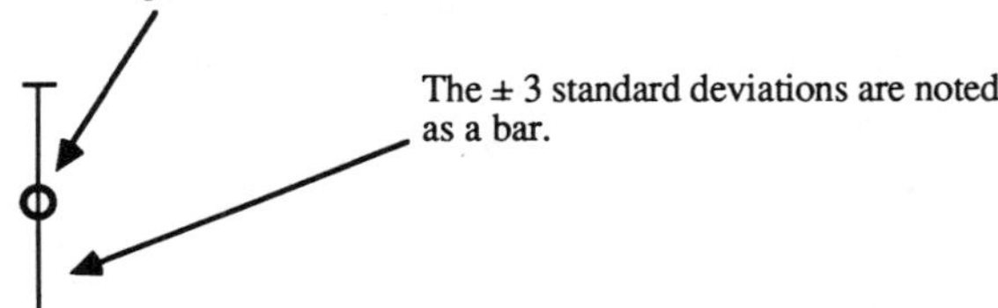

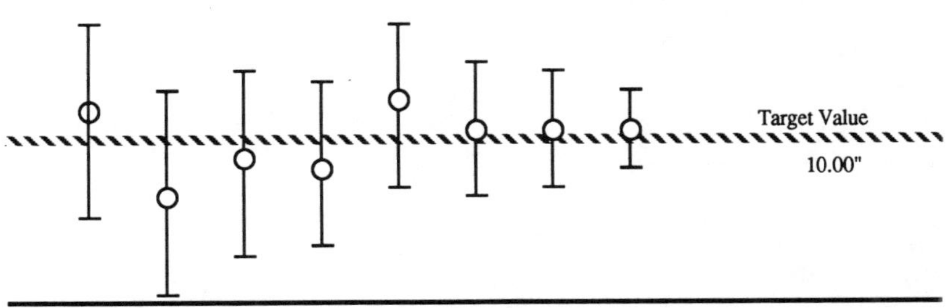

The specification for each dimension is plotted so the quality of each lot can be properly positioned. For example, the first plotting above is for a shipment that averaged 10.03" in size and had an expected variation of about 75% of the tolerance spread. Note how continuous improvement is visible as increasingly smaller part variation bars are better centered.

Figure 6.5 The Cpk chart. Used for monitoring incoming shipments or the continuous improvement of a process.

penses. Auditing is a powerful tool for verifying the effectiveness of an SPC system. It is not intended as a means of placing blame. However, the word "powerful" is all important here. An auditor with power is a problem in the making. The power must remain with the people being audited who are always seeking measures for improvement.

CHAPTER 7
SPC IN THE FRONT OFFICE

It is not unusual for some manufacturing firms to have as much as 60% of their work force not involved in the direct manufacturing of goods. If the full intention of the philosophy of continuous improvement is to be realized, then this significant part of the company must also be involved. Luckily, SPC can be readily adapted for use in an office environment, especially once SPC is a success within the manufacturing area. In addition, the manufacturing plant will make several demands for SPC related information from the front office.

SPC IN THE OFFICE ENVIRONMENT

SPC and the philosophy of continuous improvement can be used in any office setting, including manufacturers, banks, law offices, accounting firms, and other information intensive businesses. The process of using SPC is similar to its application in manufacturing. The important difference is that information is usually the product being studied. As with all SPC efforts, the first step is to adopt a strategic plan that emphasizes the principle of continuous improvement. Then problem-solving work teams are formed to study possible improvements in the information system. Their ideas are then tested against established standards.

CASE STUDY: A MANUFACTURING OFFICE AREA

In manufacturing firms there is always a front office responsible for the manufacturing support functions, such as accounting, design, sales, purchasing and so on. The Xeron Furniture Company is a manufacturing firm that also produces fasteners. Xeron employs about 500 people with about 50 employees in the front office. Xeron's top management has decided the strategic goal of the company is to produce the highest quality products at the lowest possible cost. When this goal was first drafted, management had the production area in mind. However, the same goal can easily apply to the front office as well.

The office manager, Sally, decides the problem-solving methods involved in SPC can also be applied to the office area. Sally points out to the office staff that information is the main product. Its accuracy and timeliness is critical to the support of customers. All departments within the company, the suppliers, and the people buying their manufactured products are the customers of this information. Sally then asks for a voluntary team to study and address office problems.

The first step is the identification of problems. In this case, the team draws on corrections reports and internal audits to create a list of problems. The second step is to select a problem to study. A Pareto chart of the time and estimated money involved with each problem reveals that mistakes with blueprints have created the most obvious and costly problems for the company. The front office team has great interest in this problem, since it is deeply involved with receiving and distributing of blueprints. In a typical situation, a new blueprint from a customer arrives in the Engineering Department. The print is sent to Sales and Quality Control for approval. Then the print is sent out to the manufacturing area for use. Sally's office team is responsible for the routing of blueprints among departments.

Next, the team studies the blueprint problem in detail using cost accounting estimates of the problems. The team finds that incorrect versions of a blueprint being on the shop floor accounted for $45,000 worth of mistakes every year. This information creates the baseline the team will use to evaluate the success of its future efforts to correct the problem.

Two charts are developed. The first is a monthly average of the number of incorrect blueprints discovered by manufacturing. Past counts indicate that an average of three of these are found every month. This information becomes part of a Pareto chart. This Pareto chart will record the number of incorrect versions discovered and the apparent reason the mistake occurred. A second chart will track the amount of money lost correcting the mistakes discovered. These two charts then form the baseline for the team.

The team now enters the fourth stage of problem-solving by studying the situation in detail to decide on potential actions to take. A flow chart reveals that the blueprints are handled by at least four different departments before they are used. In addition, a cause and effect chart created by the team has found that there is a no clearly defined system for destroying obsolete drawings or for notifying production that new drawings are on-site. In a typical case, the team discovers that a new drawing was in routing for three weeks. Production people continued to manufacture products under the older specifications because they had no knowledge that a drawing change had taken place. As a result, the team decided it would like to redesign the system for controlling incoming blueprints.

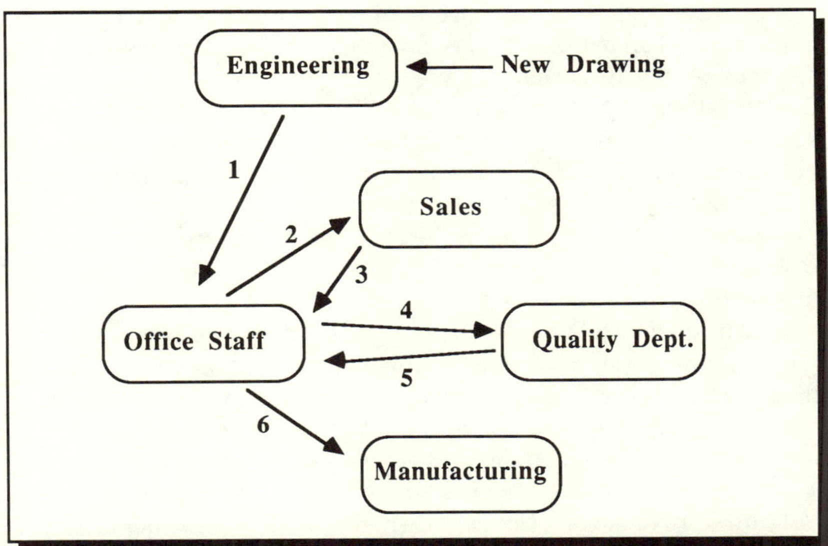

The routing of blueprints were a large problem for the office staff
at a medium-sized manufacturing firm. The problem-solving work
team created a flow chart of the routing system. The frequent
movement of the prints through the office staff seemed to created
many opportunities for lost drawings. As a solution, the system
shown below was adopted.

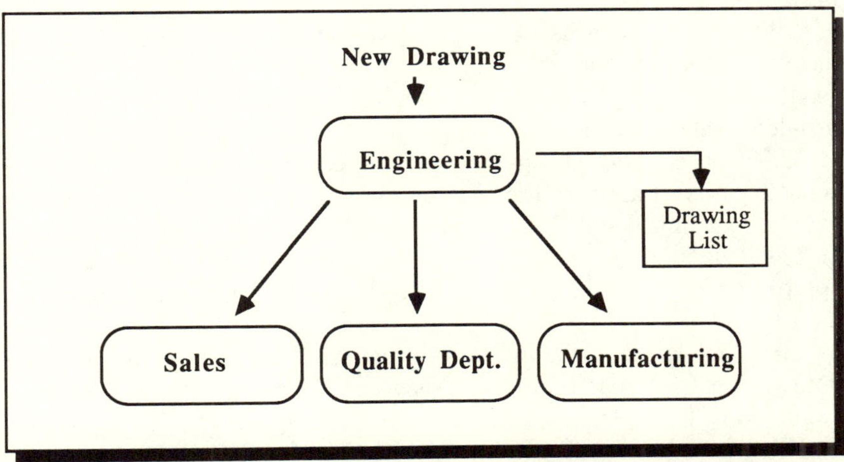

Figure 7.1 Flowcharting of a problem process.

As a fifth step, the team presents its information and ideas to top management. Top management is favorable about a new system of blueprint control, but feels that the affected department will take offense to outsiders' designing the new system. Therefore, the top managers instruct the team to call a meeting of the Sales, Engineering, Quality Control and Manufacturing Departments to make the same presentation. Top management will also be at that meeting to assign one person from each department to assist the team in the redesign of the current system. Top management also gives the newly formed group one month to report with a new system.

Action taken by the new team is the sixth step in problem-solving. In this example, the group decides that the old system was too slow and difficult. The new procedure involves the creation of a set of small forms for informing all staff of a new blueprints being on-site. In addition, one of the engineers is assigned the task of receiving all blueprints and making sure old versions are destroyed. Under the new system, a new blueprint is sent directly to Roger, the engineer. Roger fills out a short form each day with a list of new drawings that have been received. The list is copied and posted in the Sales, Quality Control and Manufacturing Departments. Roger then goes to a nearby copy shop and has four sets of the new print made. One set is sent to each affected department. Each department has five days to review the drawing. Then on every Monday morning, the four departments meet to finalize any changes to the drawing. Once completed, the new drawing is put into use and the old copies collected and destroyed by Roger.

Within two months of adopting the system, mistakes from using incorrect blueprints all but vanished. The team drew up a report on the results. In it they estimated that $40,000 was no longer spent correcting problems related to obsolete blueprints. Then they deducted the average of $2,500 in copying charges a year for copying blueprints. Thus, the economic impact appeared to be a cost reduction of $37,500. As a footnote, they also pointed out that $5,000 in mistakes were still taking place and still needed correction. And, they showed how the new system sped blueprint handling and increased communication among departments.

As a final step in the effort, the team met to discuss the continuous improvement of what they had discovered. The first possibility was obvious. John from Sales wanted to adopt the system to the sales order handling procedures. Roger introduced the idea that the entire system could be done on the office computer and then their database program could produce a monthly report of blueprint flow. This would aid in spotting future bottlenecks and forecasting heavy months for drawing corrections.

The team also identified inhibitors to the project, such as the slips falling off the posting boards and the time and expense of copying blueprints off-site. The team's effort will now focus on exploiting these opportunities and removing the inhibitors.

CASE STUDY: SPC IN A SERVICE INDUSTRY

The benefits of SPC and continuous improvement can be realized within any type of business, including service-intensive companies. For example, the banking industry is one of the largest user of these methods. The following two cases highlight how a service-intensive business can profit from the philosophies of SPC.

The Savings Bank Association of New York State reports that some of its member banks use both statistical control charts and problem-solving teams to help improve their internal operations. For example, one bank would conduct a capability study by monitoring the accuracy of new tellers. The process under study was the services provided to customers. During training, each new teller was monitored for the number of customers served and the number of mistakes made on transactions. It was found that the volume of customers would remain fairly steady but, the number of mistakes would decrease over time. Usually after 30 days the average number of mistakes became stable and predictable. For a fully trained teller, an average of two mistakes were made each day.

The bank's management understood that this stability represented statistical control. Statistical control meant that common causes of mistakes were eliminated and that random causes remained. In other words, the tellers were doing as well as they could. It was up to the management to find ways to further reduce the number of mistakes being made. In this particular case, management began an active program of working with teams of tellers to find what improvements were necessary. This resulted in regular machine maintenance and better procedures. These actions reduced the error rates for tellers even further. With fewer errors being made, the bank spent less time making corrections and talking to customers who were victims of the mistakes. This saved money and promoted better customer relations.

A different bank attacked a problem that is common in any business—customers experiencing long waits on the phone. The bank management had established a quality plan and philosophy. The major quality goal was maximum customer satisfaction. Each department formed a problem-solving team to address issues of quality that affected their department.

In one case, a support staff team decided that the largest obstacle to high customer satisfaction was in the phone answering system. Many times customers would call the bank and either be unable to get an answer or had to be transferred several times to find an available employee.

During the problem-solving team's first meeting a list of reasons why these problems occurred was compiled. Each team member had several ideas such as, people leaving their desks without forwarding their phones, inexperienced telephone operators, and busy lines. Using this list a tally sheet was drawn up for measuring the occurrence of each problem. Equipment was installed on the switchboard to count how many calls coming in received a busy signal. It was discovered that 6% of all incoming calls could not get through. The 6% formed the baseline of telephone performance.

The next step was an investigation into the actual causes of the phone delays. It was discovered that during lunch, one operator was left on the switchboard while the other operators took a lunch break. In addition, the team discovered that the operators could only find out if an employee was available was to transfer the call. If the person was tied up, the operator had to come back on the line to tell the customer to wait until someone else could be located, resulting in the familiar game of "phone tag." The team met with management to suggest two new methods of handling calls. The first was a rotation of operators and part-time clerical help at the switchboard so at least two operators were always present. The second was the instructing of other employees to either forward their phones or notify the operators when they will be unavailable. Management quickly adopted these suggestions. They were aware that 6% of their customers being unable to reach the bank by phone meant 6% less business.

A month after the new procedure was in place the team again measured the number of calls not getting through. The level had dropped to less than 1%. In addition, fewer transfers were taking place. This information was presented to management.

CONSIDERATIONS WHEN PROBLEM SOLVING IN AN OFFICE

1. Select a well-defined goal. A manufacturing team has a specific product to inspect, but the office team does not have an identifiable product so a clear-cut and measurable goal is critical to the team's success. Unlike a committee, a problem-solving team does not engage in a never

ending series of meetings. Most of the effort is placed in collecting data and acting upon it. To foster a feeling of dynamic team work, a well-defined task must always be placed before the group.

2. The critical characteristic. To increase the effectiveness of the team, part of the task must be a critical characteristic that can be measured. The first task is to describe the current situation and then, to establish a destination which must be described in measurable terms. Some possible critical characteristics for an office include:

- errors on documents
- timing of the paper flow
- absenteeism
- delays or unanswered phone calls
- math errors
- loss of documents
- response time to a customer request

3. Office work teams. An office work team will have different motivation from that of a work team in the production area. First, the office people are generally not on a productivity quota system. There are few managers emphasizing higher productivity in an office setting. In the same light, quality assurance is rarely an assigned task in the office. Instead, a group of auditors is supposed to double check work and procedures. Second, few office workers are used to real team efforts. Most have been trained to follow their manager's lead. So, if a team leader is not the highest placed official in the room, other team members will tend towards nonparticipation or will look for approval from the highest ranking person in the room before acting. And finally, increases in the efficiency of office procedure are difficult to implement and demonstrate. People resist change, especially when they are the ones who designed the original procedure.

4. Continuous Improvement. A problem that plagues management in a company is the psychology of the "daily cure." In these situations, a management group may constantly change the operating system they endorse. This year may be SPC, but tomorrow, it could be Just-in-Time. As time goes on fewer recources are dedicated to the problem-solving groups. Without the emphasis and support of the philosophy of continuous improvement, participants quickly fade from the teams to join the more popular groups.

TIES BETWEEN AN SPC SYSTEM IN A PRODUCTION AREA AND THE FRONT OFFICE.

When an SPC system is implemented in the production area, almost every department in the front office will have to participate in some way. These ties between the front office and the production area must be encouraged. One of the causes for a breakdown in an SPC system comes from the lack of cooperation between the front office and the SPC system for manufacturing. Each department in the front office can potentially gain from the information being collected by the manufacturing area.

1. Sales. The sales department is frequently the group that first encourages the implementation of SPC. The reason is simple—customers require it for new contracts. But, even without that motivation, an SPC system could greatly benefit a sales department because quality is part of the cost of a product. An example will illustrate this point.

In a sales office of an international manufacturing firm, a group of sales representatives from several smaller companies were bidding on a new contract. Each representative claimed the new part could be produced for a dollar a copy at a rate of a million a month. Each promised to deliver the parts within 24 hours of manufacture. Then the room was silent. From the back of the room one of the representatives spoke up.

"We can keep all of the produced parts to an overall length and diameter within 50% of the stated tolerance."

Naturally, the group was stunned. How could this sales representative know the level of quality of parts not yet produced? When challenged, he produced the control charts of a similar part the company had been producing for the past three years. The Cpk index on every chart was above 2.00. So, SPC can also be a powerful sales tool.

2. Purchasing. As stated previously, the SPC system will detect problems in the quality of purchased materials. Many companies have fired purchasing agents who knowingly purchased defective material. To prevent that situation, however, the same companies insist suppliers provide certification and control charts with purchased materials. In one case a steel stamping plant discovered on its control charts, that the critical dimension of every part would change dramatically when a new roll of steel was mounted. To correct the problem, the company now requires the steel mill to include information about the average thickness and variation in the steel rolls. In this way, the purchasing people can warn production departments of changes needed for deviant rolls of steel.

In another application, some purchasing agents now examine the results of precontract capability studies from suppliers. Under these cir-

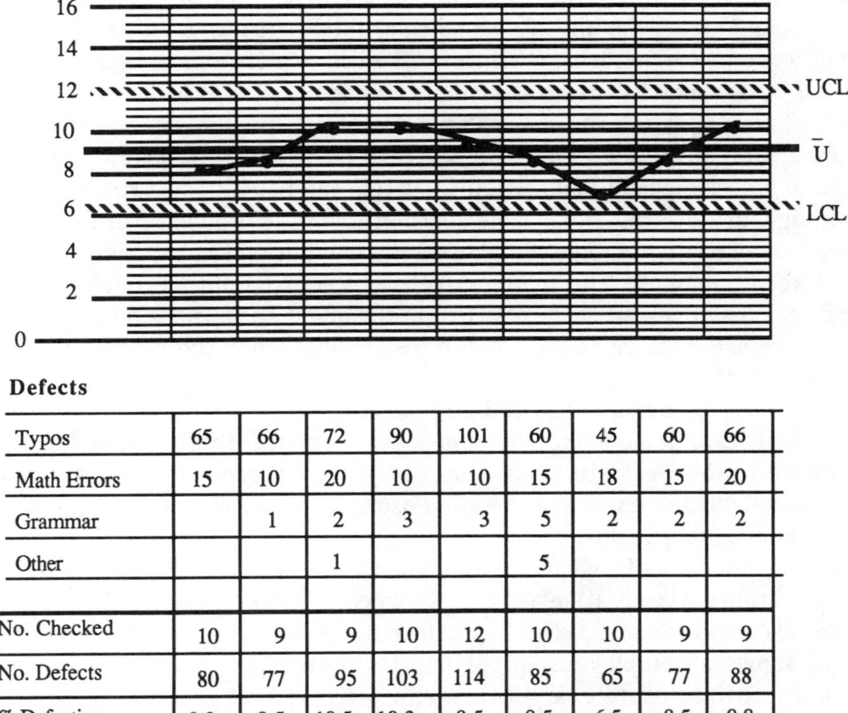

Typos	65	66	72	90	101	60	45	60	66
Math Errors	15	10	20	10	10	15	18	15	20
Grammar		1	2	3	3	5	2	2	2
Other			1			5			
No. Checked	10	9	9	10	12	10	10	9	9
No. Defects	80	77	95	103	114	85	65	77	88
% Defective	8.0	8.5	10.5	10.3	9.5	8.5	6.5	8.5	9.8

The Law office work team created a u-bar chart to monitor the number
of mistakes in office documents. Above is the resulting chart. An
average page had nine errors.

Figure 7.2 The use of a defect control chart in a non-manufacturing
environment.

cumstances a new contract can only be bid on when the supplier tries a
short run of producing the part. Thirty parts are sampled and evaluated
for average, standard deviation, and the capability of critical dimensions.
The purchaser can compare the quality and price of several bidders.

3. Engineering. The key role of engineering is the design of parts and
processes to produce the parts. SPC can benefit the engineers by providing them with timely information about the type and number of production problems encountered during manufacturing. For example, a
company that manufactures wood products may discover a problem with
warpage. By examining the SPC logs, the engineers may find a time
period when warping was nonexistent. By tracing the activities of that

time period they may uncover a fault in the design. Such a fault could be the change from the use of a wood that didn't warp to a less expensive, but warp-prone material. (Chapters 9 and 10 explore this in more detail.)

4. Customer Relations. Normally this is the group that handles complaints or requests from the customers. The job involves sorting through stacks of information to find the source of a customer complaint and suggesting possible solutions. An SPC system can benefit this group by providing more accurate quality information about the product. For instance, a group of customers has complained the handles on the company's product keep falling off. The customer relations representative could pull the lift test results about the product from the quality assurance department. If there are wide variations on life test results, then the representative can alert the manufacturing department it is critical that variations of handle strength tests need to be reduced.

The legal department would probably benefit from the same data. It is hard to demonstrate neglect on the part of a company that keeps a continual monitoring of a process and backs up production with aggressive problem-solving teams.

5. Management. Even though management has a difficult and involved role in the creation and maintenance of an SPC system, the results are also rewarding. An SPC system puts problems into hard numbers and allows all employees to participate in creating solutions, allowing managers to accurately target their efforts to the most pressing problems. At the same time, the work force takes some of the responsibility for solving the more long-term problems. In addition, a manager casually reading through SPC charts on the floor can verify the existence of a problem or the effectiveness of a solution. In short, SPC, when properly used, eliminates many of the "fire fighting" situations that a typical manager must handle.

6. Technical Support. Some manufacturing companies create products that rely upon complex production processes. The complexity of these processes requires specially trained technicians to support the production area. An example is the chemical industry. The production of chemical batches is fairly straightforward, but any corrective action or adjustments require detailed calculation. In this respect, an SPC system would include the technical support people in the reaction plans.

7. Production Planning. The control charts and process logs are invaluable to the production planning people. The information from an SPC system will detail the frequency of repairs on tools, dies and machinery which aids in the planning of production runs by forecasting downtime and lead times. In addition, the quality levels recorded in charts for

defectives and in capability studies will predict the amount of scrap a production run produces or the amount of rework to expect. This type of data is helpful in designing more just-in-time principle into the production system (see chapter 8).

8. Accounting. The accounting department will be part of the cost accounting necessary for establishing the economic impact of problems discovered by the SPC system. Such information will also help the accounting department to expand and clarify its scrap, rework and cost of quality reports to management.

9. Market Research. The market research people are responsible for finding new markets or expanding old markets for products manufactured by the company. To accomplish this task, one function this group performs is to test the market's reaction to potential or existing products. Market surveys can then be compared to the concerns being addressed by the production problem-solving teams. On one hand, they may find customers are far more interested in problems that the problem-solving teams do not address. On the other hand, they may uncover opportunities that the SPC system indicates can be exploited. Market research people may find customers are experiencing failures of electrical components in appliances. At the same time, the production problem-solving teams have devised ways of making the electrical systems on their products more reliable. The market research group can then suggest to sales that this condition be exploited.

10. Data Processing. SPC tends to create a wealth of information, but without a resource to contain, sort, and summarize it, the data is overwhelming and useless. The data processing area can be called upon to create the databases necessary to control the flood of new information, such as by adding to the production floor terminal program an area for entering information about out-of-statistical-control points. A series of sensors on the machinery could record changes in the process. Each operator could log on to a nearby computer terminal to enter job information. Thus, each production run would include information on the material used, tools involved, people on the job, costs, and the problems encountered.

11. Personnel. The use of an SPC system requires people who can perform math, read, and work with measuring standards. The personnel department can either find or train such people, but they should be trained in SPC procedures. This will help them to select the proper SPC personnel to support the system. If applicants for an SPC coordinator position are asked if they have had experience with SPC, of course, most people answer "yes," whether they have or not. A well-trained personnel employee may follow up by asking, "which type of capability study do the

applicants prefer to use?" Obviously, only those people who have worked with capability studies will understand that this question cannot be answered without making further inquires.

12. Inventory. As stated before, an SPC system is invaluable for establishing a Just-in-Time inventory system. However, SPC is also valuable to companies that are not pursuing the just-in-time philosophy. In these cases, SPC can be used to monitor the accuracy of inventory and the timeliness of deliveries. The most efficient inventory systems will make no errors in either one. An SPC problem-solving team can be used to monitor and correct any inventory problems.

13. Maintenance. The best solution to any problem is to prevent it. Preventive measures require a well-trained maintenance department. Then an SPC system is needed to accurately identify problems as they happen or just before they happen. In Japan, many plants run three shifts; the first two shifts do production work; the third shift is dedicated to preventive maintenance. These plants out produce three-shift production plants worldwide. The reason is simple—an SPC system can pinpoint processes that are gaining in variation or drifting from specification before bad parts are produced. Within one day a maintenance crew can correct the situation before it becomes a problem. In other words, the two shifts of production usually have no downtime from breakdowns or from rework. Instead, the work teams busy themselves discovering new ways for faster tool changes and higher productivity.

SUMMARY

SPC is logical in the manufacturing area, but its potential benefits to other parts of a business are often overlooked. The principle of continuous improvement was never intended solely for the production area. Any business, or individual for that matter, can benefit from the idea of improving a situation a little bit every day. The advantage of emphasizing improvement in the office area as well as production, is to further impress both customers and employees of the advantages of SPC and problem-solving. Service-intensive businesses are also finding that lessons learned on the production floor can be equally profitable in the office.

CHAPTER 8
PRODUCTION PLANNING AND SPC

Production planning is the organization of a manufacturing system to produce a product. The key assumption is that good planning will ultimately reduce the cost of manufacturing.

THE OBJECTIVES OF PRODUCTION PLANNING

To reduce the cost of manufacturing, the production planning department seeks an efficient design and effective parameters for the production process. The major objectives in production planning are:

- designing the actual product that will be produced;
- determining the best equipment to use in the manufacturing of the product;
- calculating the capacity requirements that will ultimately produce a profit for the company;
- designing the physical layout of the production process;
- designing the material handling system that will feed materials to and from the production process;
- setting the sequence of operations to be performed during production;
- establishing the time and performance requirements for the production process;
- setting the quality and quantity levels for production.

To accomplish these goals, the production planning function relies on four types of control. The first is production control, the actual day-to-day control of the production process. The second is inventory control, the regulation of the flow of materials to, from, and within the process. The third is quality control which is explored in Chapter 12 and the last is cost control, the active seeking of cost reduction methods. This means of control is emphasized in any SPC system and as such, is addressed throughout this book. Therefore, the first two methods of control will be highlighted in this chapter.

THE RELATIONSHIP OF SPC TO THE PRODUCTION PLANNING PROCESS

SPC's steady stream of data concerning the capability of manufacturing processes to meet established quality standards and the process logs that record the hour-by-hour operation of these processes are invaluable to the production planners. A complete SPC system should be an integral part of the production planning and control functions of a company. The people in these areas are responsible for the operation of the manufacturing processes. SPC is a dynamic monitoring and problem-solving technique that promotes the continuous improvement of processes. Therefore, it is only logical that the production department should be the chief user and advocate of an SPC system.

In fact, the complete SPC system should really be the keystone of a well-designed production system. Several examples will illustrate the interdependence of SPC and production.

Example 1: SPC can help establish the relationship between different stages of production.

Most manufacturing processes are the combination of production stages. A production planner may represent these stages with a flow chart that shows how one operation prepares a product for the next operation. Control charts at each stage of production can monitor those characteristics that will be critical to the next stage of production. For instance, in the manufacturing of an automobile fender, the first step is to stamp the shape of the fender out of a flat piece of steel. If the dimensions of that first stamping drift from their target values, then other operations in creating a fender will suffer. Perhaps the holes for attaching bolts will be drilled incorrectly because a reference point is now in error. Therefore, the production planner will concentrate SPC activities at that stage of production.

Example 2: SPC data used for quality and reliability standards.

The success of current tolerances is tested with capability studies and control charts. If a tolerance range or other specification is found to be difficult to maintain, then the production planners can determine if it is necessary to change production requirements. A painting specification may call for a flawless painted surface on the fender. Unfortunately, there

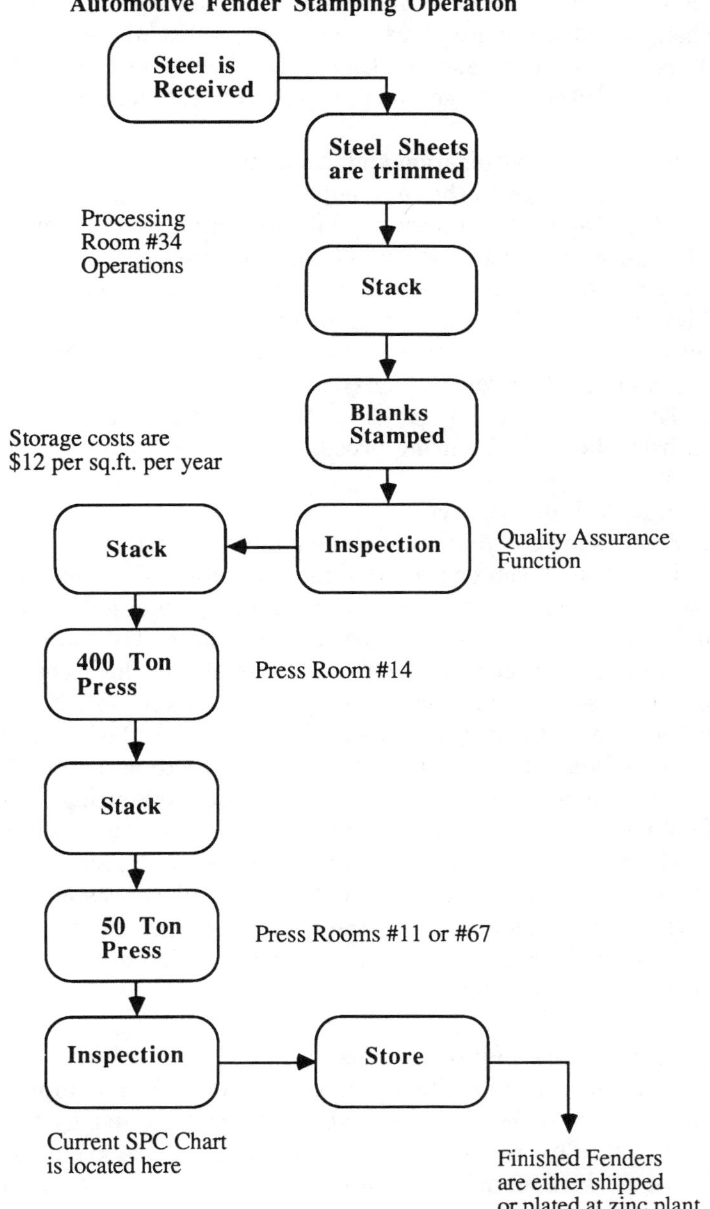

Figure 8.1 Process flowchart with notes.

is no technology that can create a flawless painted surface. The requirement must change, but attribute charts can supply a list of common defects encountered in the previous painting of the fender and thus, help the planners make intelligent decisions about a new requirement.

Example 3: Standardization and Simplification.

These are the heart of the cost reduction methods used by production planners. If a step in the process can be eliminated, production costs are reduced. A good SPC system will find such opportunities for standardizing or simplifying. A standard method for measuring a product that proves to be highly repeatable can be implemented to cut down on erroneous decisions made because quality information was formerly biased. Another case is a welding department that works as a team to eliminate weld defects. The results would include the ability to remove the weld rework function from the manufacturing process.

Example 4: Value Analysis.

An overly simple definition of value analysis is that it is the exploration of how a product can be produced more cheaply. The costs of various attributes and operations are evaluated. Then new methods and designs are tried that should reduce the cost of production. IBM did this with the keyboard of its electric typewriters. First, it switched from humans to robots for assembling the keyboard which cut expenses. Next, it examined the size and arrangement of the keys and redesigned them so that fewer motions of the robot arm were needed to assemble the keyboard. This simplified procedure reduced the cost of operating the robots.

SPC helps the process of value analysis by providing the important benchmarks of capability and expense of processes. Success cannot be claimed unless the quality of the product or procedure was improved. A process that generates defects on the typical product must be checked for the value of those defects to the company. Value can be gained by the elimination of the defects.

Example 5: Time-Motion Studies.

Time and motion studies have many applications for the improvement of a manufacturing process, but one facet of their use is deadly to an SPC system—specifically, production rates and wage quota systems. Many plants have established minimum production quotas for workers in labor contracts before an SPC system was ever considered. The result is that little or no time is left for operators to stop and monitor a production process. For an SPC system to be successful, this must change. Time requirements and quotas must reflect the need of operators to use the SPC system. Although monitoring appears to reduce a company's productivity—in the long term it is an investment in higher profit through continuous improvement. Tied to a strict quota system, most workers are not able to participate in continuous improvement. Therefore, such companies cannot compete with those who take time to improve.

Example 6: Establishing the Sequence of Operations.

Estimating the time and sequence of production stages is an important part of production planning. Sometimes control charts will detect defects caused by the sequence of operations. For example, a delicate piece of wooden furniture receives its final sanding before the finish is applied. Because finishing work is slower than sanding, a backlog of furniture waiting to be finished is created. The production planners may have anticipated the backlog by providing a staging area for sanded furniture. Unfortunately, extra handling from staging may cause scratches or dents. The discovery of these difficulties from problem-solving team work will require planners to change the sequence of operations so the sanders split their time on other projects and thus, eliminate the damage caused by staging.

Example 7: The Selection of Workstations.

A grinding operation may have dozens of similar grinders in one area. The problem for production planners is to choose the machines that will have the highest capacity and quality. The capability studies done on each machine for each part it has produced provide excellent historical profiles of machines' performances. Matched with downtime and maintenance records, the planner's job becomes almost routine.

Example 8: Analysis of Process Capacity.

The process logs provide the information about downtime and maintenance requirements, but the control chart logs reveal the hidden delays for periodic adjustments and "tweaking" of the process by operators. Tool wear and other related information is also portrayed on some control charts assisting a planner to better estimate the number and expense of work stoppages.

Example 9: Identification of Bottlenecks.

The curse of any production system is a bottleneck in the flow of products or materials in the manufacturing process. A production system can go no faster than the slowest area of production. A slower area creates a backlog of goods behind it that take up costly inventory space and at the same time slows the productivity of the areas further into the process.

Bottlenecks are usually caused by a stage of production that does not perform as well as expected in either quality or capacity. In these situations, it is handy to have well-trained problem-solving teams already in place. Causes of the bottleneck can be addressed by both the planners and the team. Reduction of these causes should then proceed more quickly and with greater efficiency.

It should go without saying that a reduction in quality standards to open up the bottleneck is no longer seen as an acceptable solution. The old procedure was to either sort out bad products, or to issue a use-as-is disposition for the lot. Neither of these methods is cost competitive and both are now actively discouraged.

Example 10: Make-or-Buy Decisions.

An interesting result of capability studies is the make-or-buy decision (see Chapter 5). A company may receive a contract to produce a shock absorber and one critical dimension of the assembly is the diameter of the dampening rod. Upon studying the machine that produces the rod, the company discovers that it cannot meet the stringent tolerances required for a successful shock absorber. At least half of all rods would have to be measured for sorting. The production planning team may choose to solve this dilemma by evaluating whether it would be cheaper to make and sort the rod, or to buy it from another company. Perhaps, a nearby precision grinding company could make the part within the tolerance.

The above illustrations demonstrate the process of production planning. The result of production planning is the creation of a *production plan*. The production plan is the design of the manufacturing system that will make a specific product. Therefore, production planning's next step will be the implementation of this plan through *production control*.

PRODUCTION CONTROL AND SPC APPLICATIONS

Production control is the actual implementation of the production plan on a day-by-day basis. It involves the following operations:

1. The development of production schedules.
2. Setting the production system into motion by issuing manufacturing orders.
3. Determining lead times.
4. Controlling and monitoring the manufacturing operations.

5. Measuring machine utilization, scrap and sources of indirect labor.
6. Controlling the production information system.

Clearly, the last three functions are almost identical to the functions of an SPC system. Control charts monitor and aid in the control of manufacturing operations measuring the utilization of equipment and materials. SPC collects a great deal of information about the production processes. The important difference is that SPC is only part of these three functions. Production control also involves methods of control and information gathering beyond the typical functions of SPC.

In Chapters 2 and 3, control charts were explained with an example of a furniture manufacturing company. The charts were used to monitor critical characteristics and to help problem-solving teams to improve the process. What was not shown was that control charts can only be successful when SPC is a well-integrated part of the production control system. The first three functions of production control come into play at this point. The production schedule must allow for charting and problem-solving activities. Manufacturing orders must include the required capability studies and the critical characteristics to measure. Lead times have to stay flexible so that faulty processes can be corrected before full production begins.

In short, a complete SPC system will be transparent within the production control system. The term transparent in this context means that an individual would have a difficult time picking out a part of the SPC system from the production control procedures. SPC must be seen as the normal way of manufacturing goods. Occurrences, such as stopping a production run to correct a point out of statistical control, must be accepted as standard operating procedure. The transformation of a company from a tightly controlled one using the production schedule to an organization using quality controls to regulate production, is a long and difficult process. However, it is not an impossible task to transform a company to the point where SPC-based production control appears as the logical way to do things. Good management support and repeated training go a long way in making the change possible (see Chapters 4 and 5).

INVENTORY CONTROL

While production control is responsible for the day-to-day operation of the manufacturing processes, inventory control is responsible for the control of the flow of materials and goods into, within, and out of those same processes. The American Production and Inventory Control Society (APICS) defines inventory control as, "the technique of maintaining stock-keeping items at desired levels, whether they be raw materials, work-in-process, or finished products." The key phrase in this definition is ". . . desired levels." Timing, amounts, and accuracy become three critical characteristics of the inventory control process and as with production processes, inventory processes can be under an SPC system using these critical characteristics. These procedures are even more important in light of the growing dependency that inventory and production control have with newer production technologies, such as Just-in-Time deliveries.

SPC as an Aid to Implementing New Inventory Control Methods: Just-In-Time (JIT)

JIT is a philosophy. Its ideal is to have a manufacturing process so well-timed that everything produced can be shipped to customers immediately, in other words, eliminating the need for product inventories. Some proponents of the philosophy even suggest that the philosophy can be applied between stages of production so no inventory space is needed in the factory. All materials and goods are either being used or moved at all times.

Several methods for pursuing the JIT philosophy, such as existing inventory software, can be redesigned for zero inventories and smaller lot sizes. An SPC system is a good introduction to JIT no matter which method is used. The work teams use the same problem-solving methods they learn for SPC, but the critical characteristic to be studied in each case is time. For example, one team in a small manufacturing firm decided to study the flow of nuts and bolts sent to its area for assembly work. It discovered that several weeks' worth of parts were stored in the warehouse. By studying the average number for each job per day, the team was able to accurately predict the nut and bolt needs a week in advance. This information was passed on to the purchasing agents and a new schedule for more frequent purchasing of smaller orders cut the inventory space to a one day supply.

For the entire company, SPC information can be invaluable for planning a JIT system. If a JIT system implies no defective products, the receivers of the products expect all components to work. Orders are for the exact number of components necessary. There is no "just-in-case" inventory allowance. SPC can help achieve flawless quality. In addition, SPC process logs will summarize the type and frequency of machine failure or preventive maintenance. This information is invaluable to the planning of JIT based production schedules. Instead of hoping for no machine failures during production, the schedule is modified so that just the right amount of time is included to handle anticipated problems.

This point is illustrated by a company that stamps out rubber gaskets for a variety of customers. The SPC team in the stamping area has found that capability to meet tolerances will last about 50,000 cycles on the tool. After that the tool has to be sharpened. One day the production planners are told to prepare a schedule for a 10,000 part run using this tool. A wise production planner will check the machine and see how many cycles the tool has gone through since the last sharpening. If it is above 40,000, then the planner must either specify that the tool be sharpened first, or the production run halted after time for sharpening.

Another important benefit of SPC for JIT is that it introduces the entire work force to the methods needed for a company-wide effort to meet a single goal. By implementing SPC, company management has practiced mass training, data gathering and problem-solving. These same skills are needed for creating a JIT environment. In addition, a successful SPC implementation often creates an increased feeling of commitment to the company by employees. Thus, people will be more comfortable with working as a team and sharing common goals.

EFFECTS OF DIFFERENT PRODUCTION CONDITIONS

No two companies produce goods the same way. However, most companies do fall into one or more of three general methods of production. Each type of production will create different views about the benefits of SPC.

1. Job Shops. These are the production processes directly geared to the customers' requests. Frequently, the orders are for small quantities of goods. A tool and die shop is the classic case of a job shop. Ongoing methods of production are rare, as each job is set up, run, and taken down in a short period of time. Therefore, the use of conventional control charts that require a continuous process is not possible. The SPC system in such situations becomes more ingrained with production and cost control systems. Information is collected on set-up times, tool history, and common problems in obtaining stock, et cetera. The role of problem-solving teams is to then examine ways to increase turnaround times, reduce scrap and improve the final quality of production.

A tool and die shop may choose to study tool histories by job and person using them. Placing the average time until repair or replacement

Tool in Use

Job	Person	1123	375	419	113	1122	212
A -1234	John			L			
	Mary						
	Dick	X			Sh		
	Norman		L				
	Alice			X			
B-4321	Sally	Sh				X	
	Bob	Sh					
	Norman	Sh				X	

Key

☐ = No change

X = Tool broke in use

Sh = Tool needed sharpening

L = Tool lost

Note how tool 1123 requires frequent sharpening. Some jobs seem to break tools. Some people seem to lose tools.

Figure 8.2 Tool history and defect tracking grid.

on a grid, the teams can pinpoint particularly troublesome tools. In addition, they may discover people or procedures that extend the useful life of the tools. An idea could be as simple as cleaning the tools, or as complex as advanced experimental designs to determine ideal feed and speed rates for cutting tools.

2. Mass Production. Mass production occurs in companies that produce large numbers of parts, components, or final products for stock. The

manufacturing process tends to run a single item for many hours or days. After that, the process may be slightly modified to run another item. Automotive supply companies are typical of this kind of operation. Parts are produced in large numbers for specific customers. This is also the classic location for control charts and a complete SPC system. The process can be stopped from time to time and individual products examined.

3. Continuous Production. A manufacturing process that cannot be stopped once the production begins is a continuous production. Chemical companies are just one example. Once chemical mixing and processing begin, they continue until completion. Monitoring and correcting must be done on the fly. In addition, sampling of the process may take hours to analyze and require a chemist to effect a corrective action. Situations such as these can use an SPC system to monitor process variables such as temperature or pressure. However, a regular system of experimentation is required for continuous improvement.

A foundry is another good illustration of this situation. In a foundry that melts iron for castings, the pig iron is melted down in the furnace, sampled, and then poured into molds. The sample usually tests the chemistry of the iron. Things such as percent carbon and crystal formation are examined. Unfortunately, by the time the analysis is done, the metal has been poured. Sometimes, a quick carbon check will help the metallurgist correct the batch. In either case, a control chart is of little help. What is needed is a precise control on the amount of material being fed into the furnace and the temperatures being applied. By studying a history of these, the metallurgist can analyze the sensitivity of the chemistry of the iron to the factors the smelters can control. More precise melting procedures can then be developed.

COMPANY SIZE

Another factor affecting the use of production planning and SPC is the size of a company. A shop with fewer than 100 employees may find that one person is the production planning and control person. Usually, this person is also the production manager. Sometimes, a single engineer has this role. In either case, the small company has to either train this person to high skill levels or hire an assistant, possibly a consultant. Planning is a time consuming occupation, and other responsibilities tend to take priority.

Larger companies face different problems because they tend to have production planning, production control, inventory control, and SPC functions divided into separate departments. The coordination of these departments and the need to work as a team is critical to improvement. However, long held political and social divisions among departments can be difficult to overcome. Larger companies need a strong motivating force originating with upper management. Training together and providing rewards only available from interdepartmental cooperation are just two methods for breaking down the barriers among departments (see Chapter 5).

SUMMARY

Production technology is the study of joining human skills, machines, and materials to create a product. Production planning is the action a company takes to decide which methods will be used to manufacture its products. Production planning personnel set up the operating procedures to follow during manufacturing and the requirements of the production process. Controls on production, inventory, quality, and costs are needed to make the production planning process work. All of these types of controls can benefit from a complete SPC system. Control charts, capability studies, and process logs provide information for the decisions the production planners must make. However, the benefits are greatest when the SPC system is transparent to the people using the system. SPC cannot stand alone. It must become the system and not be an identifiable part added on to an older system.

CHAPTER 9
ENGINEERING AND SPC

Currently, American business is undergoing massive changes in management techniques, which tend to mask the revolution that is also taking place in engineering. Engineering is in the mainstream of implementing new manufacturing technologies and these new methods of manufacturing require a high level of technical expertise in several fields, such as statistics, mechanics, programming, et cetera. This phenomenon has resulted in engineers being chosen as resource people for new methods. Engineers are the main designers of faster, cheaper, and higher quality production systems. Therefore, they have two primary functions: the creation of designs and procedures for new products, and the development of new methods of manufacturing to keep the company competitive.

Consequently, the engineering department and the SPC system will be very dependent on one another. Engineers will need information from the SPC system for planning, designing and coordinating current production. Engineers also need to know the strengths and weaknesses of the current production methods to help them plan more efficient techniques. In addition, the SPC system requires the engineers' specifications of processes and products to function effectively. Once these relationships are developed, the company can more effectively implement other new manufacturing technologies and engineering methods.

PRODUCT DEVELOPMENT AND ENGINEERING

The first function of most engineering departments is project development, which involves either the planning of a new product or the review of an existing product. Engineering's role is to create a design that will produce a product of high quality and low manufacturing costs. To accomplish this objective, a formal procedure is followed in the development of a product and it is frequently referred to as "design review." The following steps are typical for the design review process:

1. A new product is asked for, or an existing product is nominated for review.

2. Research is done on the current world standards for this type of product and could include topics such as quality, safety, legal obligations, reliability, and market expectations. Most departments within a company participate in this part of the review by collecting information and submitting it to the design review group.
3. Using the above information, the engineers draw up specifications for the product and the processes that will produce the product.
4. With these specifications, a preliminary design is created or the existing design modified.
5. Engineering presents the new design to other departments so they can test and evaluate the proposed product. For example, sales might evaluate the marketability of the product while manufacturing would address potential problems in producing the product.
6. At a meeting of all involved departments, the final design of the product is worked out. During this meeting, the final cost estimates are also calculated.
7. Manufacturing and shipping procedures are finalized, including the methods for controlling the gages and tools used to make the product.
8. Production begins. The SPC-system will supply the first feedback on the manufacturability of the product.

After production begins, the SPC problem-solving teams will work closely with engineering to help monitor and later improve the manufacturability of the product. Good design practices will focus on eliminating as many potential problems from the product as possible, which is the first step in making the product competitive. The fewer design changes that take place after start-up, the greater the probability of higher quality. The second step is the reduction of manufacturing costs. A close working relationship between these two groups is necessary for wide scale cost reduction efforts. There are five major functions of engineering that must be actively integrated into the SPC system to promote effective cost reductions during the design review process, and later during production:

1. Information feedback during design.
2. Setting specifications.
3. Reliability testing.
4. Gage control before manufacturing.
5. Tool control during manufacturing.

INFORMATION FEEDBACK DURING DESIGN

It takes competent, objective, and experienced people to do design work who must work actively with other departments to gather information about the requirements for a product and who must employ good judgment in constantly challenging a design. The design must be examined from as many angles as possible, such as, performance, cost of manufacturing, and cost of maintenance. Therefore, engineers need information from a variety of sources and the SPC system will be a major one.

SPC supplies information for determining the reliability, maintainability, safety, and manufacturability of a product. For example, previous capability studies and control charts have tested the tools, materials, methods, and people in the production process. Also, by reading control charts and process logs, the engineer can evaluate the success of all of these elements together. This procedure will aid in the creation of specifications for manufacturing capacity, rate of production, and expected levels of quality.

An engineer is trying to determine how many bookshelves a day the Xeron Furniture Company should produce. The design in question is for a new type of bookshelf that features glass instead of wood shelves. The glass was chosen to eliminate the warping problems experienced with wood shelves. However, the question remains as to how many of these new types of bookshelves can the bookcase team make every eight hours? The engineer knows how long it takes to assemble the outer case from previous bookcase line records; the problem is determining the time to cut slots into the case for the shelves. The bookcase team has never had to make such cuts before. The office desk team's records provide the solution. This team regularly has to cut grooves in the desks and process logs reveal the time needed to train operators, how long it takes to cut the wood, and what types of problems are experienced while making the cuts.

SETTING ENGINEERING SPECIFICATIONS

In most production systems, engineers carry the responsibility of designating specification for both products and processes. In addition, customer demands, federal regulations, lawsuits, and economic pressure to produce higher quality goods have increased the importance of accurate specifications and systems to confirm their application. All of these factors mean an engineer needs far more information on the success of specifications in use then at any other time in recent history.

The SPC system can provide much of the information the engineering department needs to make better decisions. A properly designed SPC system can be used to help sort out several common situations associated with any process.

1. The effect of machine settings on specifications.

In the case of a plastic blow molding process, the operators may be attempting to mold a small bottle. The dimensions of the bottle are vital to its usefulness. A control chart could track one or two critical dimensions while a process log records the changes made in machine settings. Each time a dimension loses statistical control, the operator should note the most recent change in settings. This information is then passed to the engineers so they can begin to form a picture of the effect of various machine settings on dimensions.

2. The appropriateness of the specifications can be evaluated.

Some traditional styles of engineering dictated that once a specification was set, it was unchangeable. The assumption was that the specification was appropriate if engineering deemed it so. However, the realities of manufacturing usually include situations where the exact specification is difficult to determine. One story in engineering circles involves a young apprentice engineer calling the boss and asking what tolerance to place on a part's dimension. The supervising engineer responded by telling the young engineer to look up the tolerance on a similar part and then reduce it by 10%.

Once control charts are placed in the manufacturing area, one of the most common discoveries is that some of the processes are not capable of meeting the specifications set by engineering. The first reaction to this situation is always to try and find ways of improving the process. However, the second course of action is to address the appropriateness of a specification. Sometimes this involves an examination of the role of the dimension on the products' function. For example, a hole drilled to half an inch, plus or minus 0.002 of an inch, could be mated to a bolt that is a full 0.1 of an inch more narrow than the hole. Therefore, the tolerance range of 0.004 of an inch makes little sense.

3. One-sided specifications.

Another problem with specifications involves one-sided tolerances. A capability study tells the manufacturing people how much variation to expect on both sides of an optimal average. This leads to a problem when the specification gives an optimal value with a tolerance range on only one side of this value. A blueprint may designate that a piece of metal must be cut 10 inches in length with a tolerance range of 0 to plus 0.10 inch. The manufacturing people will be confused by this specification. They know that a process will distribute sizes on both sides of an average.

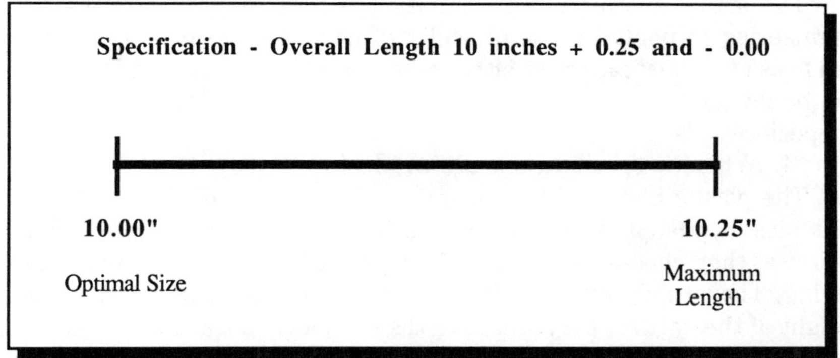

A typical one-sided tolerance from a blue print. The question for the process operator is how to set the process to best meet this specification.

Option 1: Set process average at the optimal setting so as many pieces as possible are produced at ten inches. Results - half of lot is too small.

Option 2: Set process so no pieces are too short. Results - no ideal pieces.

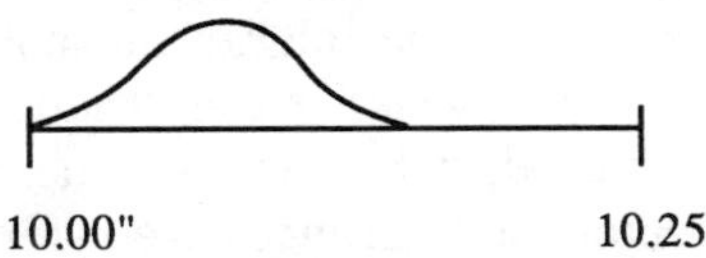

Option 3: Set process so pieces are as long as allowable. Results - all need rework.

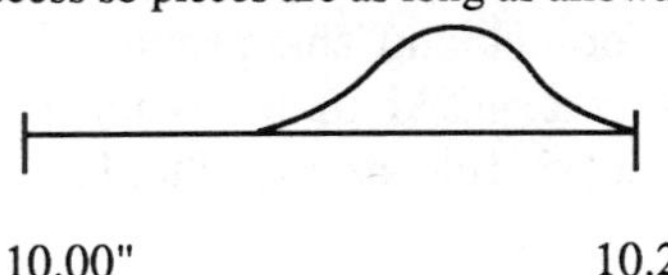

Figure 9.1 Some problems with a one-sided tolerance.

Should they set the cut to 10.05 inches so all parts are in specification, with almost none at the optimal? Should the cut be set at ten inches to produce the most cuts at the optimal size and then require half to be thrown away as too small? Perhaps, the specification is saying that longer pieces are better, so the process should be set at 10.10 inches for maximum effect.

The actual answer to this situation is to arrange a meeting with engineering to point out the difficulty of using a one-sided specification and to ask for clarification. In either case, it is certain that the SPC system can be invaluable, and sometimes annoying, in highlighting weaknesses in specifications.

4. When no specifications are available.

The plastic blow molding operators from the example above know that plastic wasted is money wasted. Therefore, as a problem-solving exercise, they choose to monitor the weight of the bottles they are producing. Their thinking is that if they can stabilize or reduce the average weight of the bottles, they will save the company thousands of dollars in plastic costs. In order to accomplish this they need to know the optimal weight for the bottles and the tolerance of the weight to help calculate Cpk indexes. They may discover that no such specification was ever developed. The current specification may only say that enough plastic should be used to create a complete bottle, so the engineers can assist the problem-solving team by determining the optimal weights for the bottles produced. Then the team can proceed to make control charts and adjustments based on this additional information.

Note on Geometric Tolerancing

There is another solution for some of the problems listed above. The use and interpretation of tolerances can be greatly clarified with the aid of geometric tolerancing. Geometric tolerancing and dimensioning is a common design language for describing the "shape" of an object on a blueprint. When correctly used, it has uniform meaning to any trained engineer. In addition, geometric tolerances are referenced to other surfaces or points on the part. In this way, it is easy to understand the desired dimensions of an object and the proper method for confirming these dimensions. Those interested in investigating the problem preventing potentials of geometric tolerancing should refer to ANSI Y14.5–1982.

RELIABILITY TESTING

After the design is created and the specifications noted, the next step for engineering is to test the proposed product. Reliability is the testing of the product's performance over time. Specifically, the goal of reliability testing is to establish the expected performance in a stated environment over a given period of time. Usually, computer modeling, prototypes or

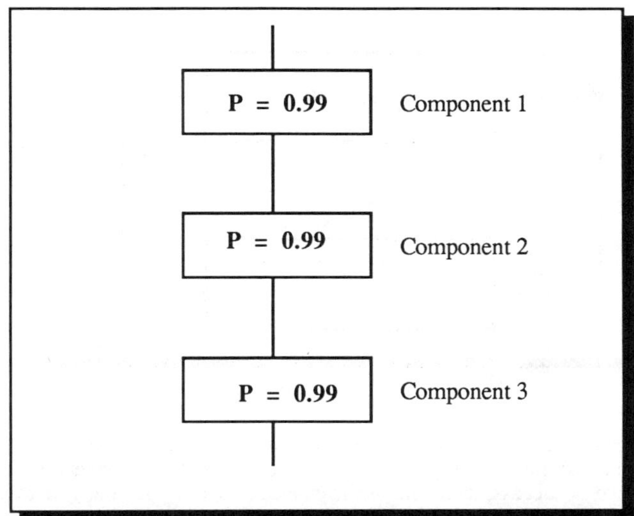

Three components are assembled into one system. For the system to work, all three components must work (connected in series). Each component has a 99% chance of success. The probability of the system working is expressed by multiplying the component probabilities together.

$$P = 0.99 \times 0.99 \times 0.99 = 0.970299$$

Thus the chance of success is just over 97%, while the chance of failure is just under 3% (1.00 - P).

Figure 9.2 The probability of failure in a system in series.

mock-ups are used. This information helps the sales, marketing, and warranty people plan how the product will be presented to the buying public.

An electrical circuit design can illustrate the exchange of information between SPC and reliability testing. To begin with, the resistors, transitors, diodes, and other parts of the circuit have probably been used in other designs. The known reliability of each part can be used to calculate the probable success of the proposed component. The engineer may note that three particular resistors will be used in series in the circuit. By checking past success ratios for these resistors from the capability studies, he finds that each has a 99% chance of working correctly for one year.

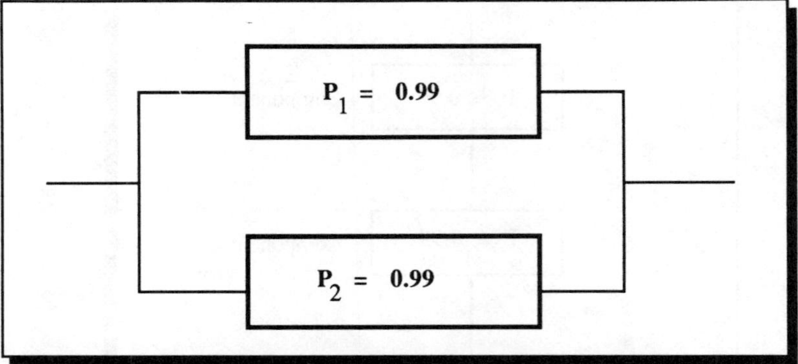

Suppose that two components are used in redundancy. If one component should fail, then the other would take over. If both units have a 99% chance of success, then what is the chance of success for the system in redundancy? The following formula is used to calculate the answer.

$$\text{Realibility} = 1 - (1 - P)^n$$

In other words,

$$R = 1 - (1 - 0.99)(1 - 0.99)$$

$$R = 1 - (0.01 \times 0.01)$$

$$R = 0.9999$$

or, a 99.99% chance for success.

Figure 9.3 The probability of success for a system connected in redundancy.

A simple probability calculation indicates that three of these resistors in series have a 97% chance of success. In other words, the manufacturer can expect an average 3% of the circuits failing during the first year of use.

Armed with these estimates, the engineer can take countermeasures against weak areas of the design. The three resistors may be replaced by a single, more reliable component, or a higher quality resistor may be specified. If resistors with 99.9% success rates were used, the series would produce about 0.3% defective circuits. However, if a parallel circuit were designed with other resistors added as redundancy and all resistors still had the 99% success ratio, then only about 0.00001% of the circuits should fail. A simple calculation of cost savings would indicate the best countermeasure to employ.

Another use of SPC data for reliability engineering is the record of failure rates for the history of a product. Upon start-up and during early production, there may be a high rate of parts failing. As corrective actions are implemented, these failure rates would fall until acceptable levels were reached (zero being ideal). By matching these changing failure rates to life history curves, engineers can estimate how long corrective actions and additional personnel will be needed to bring a new product up to specified quality levels.

Most of the work of reliability engineering involves analysis of the product while it is in use. Problems reported by customers or by extended life testing are used to modify future designs of the product. Information from field use and testing can be valuable for the SPC problem-solving teams and can be used to select critical problems to address during production. The plastic bottle maker may find that many bottles tend to crack easily after sitting on shelves for more than a year. The plastic blow mold team would then work with engineering to test potentially stronger mixes of plastic.

GAGE CONTROL BEFORE MANUFACTURING

Two aspects of measurement gages need to be evaluated on a regular basis. The first is the calibration of each unit. Every device that measures parts should be sent to a professional lab for calibration at least once a year. In addition, each gage should have a sticker attached to it indicating when the next calibration is due. A separate log sheet of calibration should be on file for every gage in use.

Calibration will confirm that a given gage will accurately measure an established standard. Another concern is whether the gage will work accurately when measuring a specific part on the shop floor. The test of the gage's ability to work in the factory environment is called the Index of Repeatability and Reproducibility (R&R for short).

A good R&R study will check for two possible sources of error. The first is called the *Fixture Variation*. This is the error encountered when a gage measures the same dimension repeatedly on the same part. There will always be slight differences in readings which is also called the *Repeatability* of the gage.

The second source of error is called *Appraiser Variation* that is the differences in readings taken on the same part by different operators of the gage. An example of this is the use of a single caliper by an inspector on each of the three shifts. This is also called the *Reproducibility* of the gage.

The amount of R&R error must not be more than 10% of the tolerance given for the dimension measured. The same holds true for the appraiser and fixture variation. A fixture variation too high could mean the gage needs repair or that its precision is too low for the specific application. An appraiser variation that is too high could mean the gage operators should have training in the use of the gage or that measuring standards should be established. Appendix B has a complete description of how to conduct the R&R study.

TOOL CONTROL DURING MANUFACTURING

Tool management is a simple idea that is becoming more difficult to implement every day. To begin with, the basic intention of tool management is to supply the right tool or gage to the proper machine, on time. Problems begin when a company also wants to monitor the performance of tools and decrease set-up times. Therefore, the tool crib personnel need to be able to track the physical location of tooling and its current condition. No longer can an operator go to a tool crib and pick up the necessary tools just by asking for them. Now engineering has to prepare bills of material just for the tooling required for a product.

Luckily, computers and SPC can help in this situation. First of all, computers can speed the storage and retrieval of tool information. Tools can be marked with bar codes to help the tool crib people to track who has checked out which tools. Second, the control charts and capability studies are helpful for predicting the times between repair or replacement for tools. The R&R studies of gages provide information on the usability of measuring equipment.

All of the above information is fed to a central computer for the tool cribs. At each crib, the tool manager now has all information needed to run a smooth operation. The computer can summarize the bill of materials scheduled for use each day so that tools can be located and queued up for immediate use. It would also be possible to create a daily report of which tools would soon be scrapped or repaired that would assist in predicting when new tools should be ordered. By tracking the availability of selected tools in advance, each crib would use only the exact number of tools required for current production. Such an ideal situation would cut costs from lost tools, unnecessary repairs and needless delays.

Engineering's chief role in this process is the design of this tool tracking system and the analysis of the tool histories it would create. If some tools wear out or break frequently, then changes in tool specifications or manufacturing procedures may be necessary. In addition, engineering would work with the SPC teams to help design tools that could be changed more quickly, cutting lead times and thus, reduce production expenses.

A story can illustrate the importance of good tool management when using SPC and other new technologies. The new owner of a metal stamping plant began examining the current condition of the stamping dies and was surprised that the plant manager had a very difficult time locating even half of the known dies. Further searching revealed that several of the dies were sitting in the scrap yard waiting to be junked. Unfortunately, many of them had never been used. There were no tool histories kept and no tracking system, so perfectly good dies were mixed with worn out ones. Needless to say, many of the control charts were detecting faulty processes because no one had information about the current conditions of a die in use.

THE ROAD TO NEW MANUFACTURING TECHNOLOGIES

A person not familiar with manufacturing operations may believe that if a company installs robots and computers into its process, the quality and productivity of the plant will automatically increase. In fact, nothing could be further from the truth. Several companies have recently discovered that new technologies can actually hurt productivity. The computer is an excellent example. If a company has been operating with an accounting system that makes frequent mistakes, a computer added to the system will only speed up the rate at which the mistakes are made. Thus, the golden rule of implementing new technologies is, "automation magnifies the effect of the system it is applied to." That is, automation will magnify the effect of a bad system to make it worse. And, in all fairness, a good system will be magnified into a great system.

Below are examples of some of the new technologies available for manufacturers. The best way to create a disaster with these technologies is to implement them in ignorance. This ignorance can be dual in nature. The first type is ignorance of the technology being implemented. Without familiarity of the potentials available, the wrong equipment and software could be purchased or the materials might not be implemented for maximum benefit. Chapter 13 will examine these points in detail. The second danger is ignorance of the processes where the technology will be implemented. For example, automation of a process will probably not change its capability so faulty processes will continue to produce low quality products.

What the SPC system does is find the problems and opportunities that have to be addressed before a new technology is used. Some situations will be both a problem and an opportunity. For example, the study of a grinding machine with an SPC system reveals that without constant adjustment of the grinding wheels, the tolerances can not be met. Therefore, the group planning to automate the grinding process decides to add one more feature, a laser measurement device that will measure each part during grinding and automatically adjust the grinding wheels until an optimal size is reached. In addition, the computer controlling the laser and grinding wheels will also monitor the amount of grinding surface left so that wheels can be changed just as they wear out. The added knowledge of problems in the process found by the SPC system can be invaluable to the design of automated processes.

COMPUTER ASSISTED MANUFACTURING

Computers can play dozens of different roles in assisting manufacturing processes. Each of these roles can be enhanced with the information provided form an SPC system. Several examples follow.

CAD/CAM—Computer Aided Design (CAD) is the use of a computer in the drawing, testing, and describing of a product's design. In short, a computer replaces the hand drawn activities normally associated with design. Computer Aided Manufacturing (CAM) is the use of the electronic designs for the creation of machine instructions to create the product. The classic example is one of a part drawn on a computer screen and then translated to machine instructions for a Computer Numerical Control (CNC) machine to produce. At no time does a physical drawing or prototype exist. This capability speeds engineering time by reducing the number of steps that require humans.

The full potential of CAD/CAM is not realized unless they become part of the SPC system. Engineers cannot operate in a vacuum. They must know how manufacturable a product is. The SPC system must supply the engineers with information such as, problems being experienced with current tools, fixtures, and dies. Then, engineers can form problem-

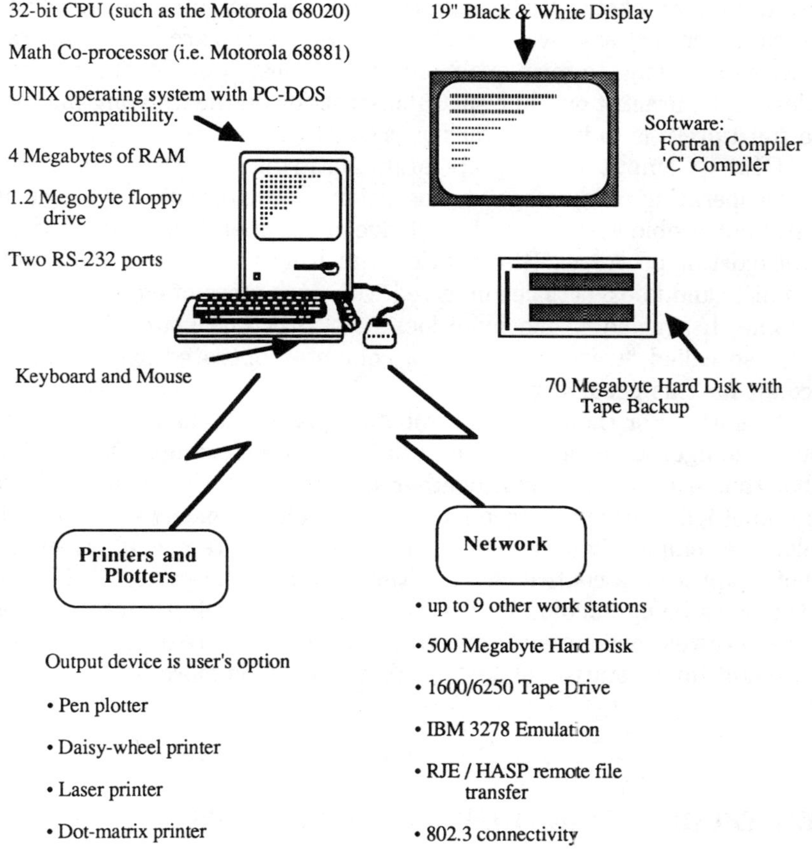

Figure 9.4 A microcomputer based engineering workstation.

solving teams with production workers to help reduce these problems. The advantage of having CAD/CAM is that new design ideas suggested by production can be quickly drawn, modeled, and tested on a computer which reduces the cost of trying new ideas, and increases the number and type of ideas that can be tested.

CAD software packages are increasingly adding database capabilities—that is, the ability to store information about parts of the drawing in a separate area. SPC can help provide production feedback for future designs. A stamping plant engineer may be designing a new bracket with a 90 degree bend in a thin piece of steel. Using the CAD workstation, this design engineer calls up the database for all previously designed brackets. The computer is instructed to read all SPC chart notes submitted about brackets with 90 degree bends. The computer quickly locates several notes that say two of the three main presses are known to crack metal while trying to form such angles. The design engineer can either redesign the bracket or add a special instruction on the drawing that only the third press is to be used during production.

CIM—Computer Integrate Manufacturing (CIM) involves computers operating every stage of production. These proposed systems are nearly impossible without the knowledge of the operating characteristics of the existing processes. SPC provides much of this information including the flukes and idiosyncrasies involved with each piece of equipment. For example, if a grinding machine does best after one hour of producing parts, so called "warm up", then a computer operated system has to account for this peculiarity.

Robots—For the most part, robots replace humans for the monotonous or dangerous handling of materials. A classic example is one of the robot that places two parts together for an automatic welder to weld. The robot will continue to feed parts to the welding machine whether the welds are completely successful or not. An SPC system would assist the robot by putting a control chart at the source of the parts. If the sizes of the parts being handled by the robot vary too much, then the success of the autoweld will suffer. However, if the process creating these parts is brought under statistical control, then success is more likely.

NEW TECHNOLOGIES DISPLACING WORKERS

The news media have reported frequently on the impact of new technologies on jobs. The usual situation is that automation has eliminated the need for several types of jobs. Some companies lay off workers who used to operate the manual processes. However, this is not encouraged by the principles of SPC and continuous improvement. Instead, displaced workers should be shifted to positions that further assure the quality of the products and the processes. An example is the use of CNC machinery. A CNC machine requires less involvement time from the operator during production runs. This additional time could be put to use to study the capabilities of the process, to participate in problem-solving teams, or to learn more about the uses of the CNC controller in class. In short, people are too valuable a resource to dispose of when a new technology reduces their workload. Instead, people can be the vital force in an SPC system.

SUMMARY

Any new technology is best implemented when the process it will complement is well understood. The use of specifications for products and processes works best when effectiveness is evaluated regularly. Both of these requirements are served by a complete SPC system, which furnishes timely information about the success of specifications in use within given processes. Control charts and process capability studies can provide a wealth of knowledge about the effectiveness of current specifications. In addition, they pinpoint the trouble spots that new technologies must address. Just as knowledge is valuable for development and improvement, it also increases competition. When people and technologies work in harmonious teams, productivity increases exponentially.

CHAPTER 10
DESIGN AS THE ULTIMATE COMPLEMENT TO AN SPC SYSTEM

The ultimate source of high product quality and process efficiency is good design. Proper design methods begin with an examination of the information provided by a complete SPC system. They end using this information to conduct experiments on the process and its resulting products. The goal of such experiments is to take obvious information (the normal behavior of the production process) and use it to find what lies just beyond the obvious, such as optimal settings, better tolerances, or more cost effective products. Such experimentation has traditionally been expensive. However, Dr. Taguchi has revitalized the interest in experimentation with the creation of cost effective methods of research and development.

In the past, scientific experimentation with processes and product designs was associated with the research and development department. However, new technologies have been introduced that spread the use of experimental design to all parts of the factory. The computer is one such technology that allows an experimenter to analyze thousands of results in just a few minutes. Another technology is Taguchi's methods of experimental design. Taguchi's methods are a shortcut in the experimental process that significantly reduce the cost of conducting experiments.

This chapter will overview the use of experimentation in improving designs. Emphasis will be placed on the newer Taguchi methods. The following material is an overview of the potentials available with the newer methods. Hundreds of hours of training and practice are required to master these methods. Experimentation should not be conducted casually.

THE OLD METHOD: CLASSIC EXPERIMENTAL DESIGN

The purpose of any experiment is to find evidence of selected factors having a predictable effect on a given situation. A single cause-and-effect situation can illustrate this idea. If an egg is struck with a hammer it will shatter. The energy of the moving hammer is the cause of the effect, the egg shattering. The hammer is a *factor* and the shattered egg is a *result*. This is an obvious cause-and-effect situation. The experiment is intended to investigate those situations that are less obvious than the egg and the hammer. These investigations always begin with a question, such as "If the plastic molding machine is used in a room where the temperature changes frequently, will this affect the final product?"

An experiment can also find patterns of interaction. If an egg were heated in a microwave oven and then hit with a hammer, the egg would shatter much more dramatically. This is the result of two or more factors interacting with each other. In this case, the steam in the egg combined with the energy in the moving hammer to create a more devastating result. The experimenters on the plastic molding machine may also be interested in interactions; they may ask, "Would room temperature and mold temperature variations have a combined effect on product quality?"

The method for a classic experiment is to form a hypothesis from each of the questions posed by the experimenters. Usually, the hypothesis is stated in the null, as if no cause-and-effect were the normal situation. Such a null hypothesis may be spelled out as "Variations in the room temperature have no effect on the final dimensions of the products produced by the plastic molding machine." The classical method of experimentation dictates that the experimenter now must try to confirm this hypothesis with empirical evidence. The final step of the experiment will be to either accept or reject the null hypothesis.

To find this evidence, the experimenter designs an experiment to test the hypothesis. There are many forms of experimental designs. The classic design is to create two groups. The first group is called a "control group" and will remain unchanged during the experiment. The other group is called the "treatment group" and will experience changes in selected factors. The two groups will be compared for differences. In the plastic molding example, one machine will be isolated from variations in room temperatures (control). The other machine will be subjected to the normal variations in room temperatures (treatment). All other factors such as products being produced, machine settings, and the plastic material being used will be identical on both machines. The experiment will involve running some production and then measuring the variations in dimensions for products from both machines. If no significant variation is found between dimensions for parts from both machines, then there is evidence for the null hypothesis. However, if the machine used in different room temperatures makes products with higher variations in dimensions, then the null hypothesis is rejected.

The engineer for a plastic molding shop decides to experiment with
molding machines. He decides to test four factors, one at a time, for their
effect on a serious problem -- shrinkage. The four factors are mold temperature,
pre-molding plastic temperature, screw speed, and moisture. Each factor in
a classic experiment would be tested -- one at a time.

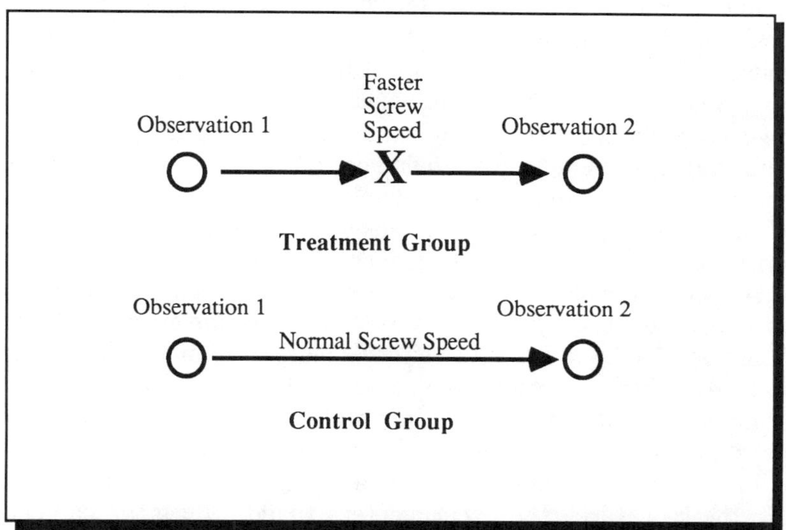

In a classic experiment the process is run under two conditions. The first is
the normal operation of the machine. This time is called the "control group of
parts." The effect on these parts is measured. Then only one factor is
changed in this case a faster screw speed is used on the machine. These
parts are called the "treatment group." If they differ significantly from the
control group, then the effect of the treatment is seen as significant.

The test for a significant difference is expressed as a Null Hypothesis.
For example, the null hypothesis for the above example would read,
"Changing the screw speed of the molding machine has no effect on the
shrinkage of parts." In other words,

$$\text{Treatment Group} = \text{Control Group}$$

or,

The shrinkage in the Treatment Group = Shrinkage in the Control Group

If the shrinkage in both groups seems to be statistically similar, then the null
hypothesis is accepted -- screw speed seems to have no effect on shrinkage.

Figure 10.1 The Classic Experimental Design—One Factor at a Time.

This method of experimental design is simple and effective for finding differences caused by one factor. Unfortunately, the industrial environment will expose a process or a product to hundreds of factors that are varying continuously. In reality, it is impossible for the production department to control more than a few factors at one time. Therefore, a different method of experimentation is needed to study many factors at the same time to find the few that must be controlled. Luckily, *factorial experiments* can be conducted to evaluate the simultaneous effect of a set of factors on one result. For example, the settings on the plastic molding machine can be studied to find out which combination of settings will produce the best quality in the final product.

Unfortunately, factorial designs also show up the weaknesses of using experimental designs on the factory floor. The plastic molding machine is a good example. the plastic molding machine has thirteen possible controls that the operator can adjust. If an experiment is designed to test each combination of the controls set to either a high, medium or a low setting, almost 1.6 million experiments would be required since the machine settings for thirteen controls set either high, medium or low can create almost 1.6 million combinations. Each combination would have to be tried to find the single best combination of settings.

In the real world, the experimenters would reduce the number of settings to study and select the ones most likely to change the process by studying the control logs. The logs would document which controls were adjusted whenever the process lost statistical control. They may select screw speed, premolding temperature, and mold pressure as the only three factors for study. Then only eight combinations have to be tried. The results are interpreted by using effect diagrams that show the change in final product quality (such as total shrinkage) for each of the settings. Using these charts, they select the setting that minimizes shrinkage. There is some danger from not knowing the effect of the other ten settings on these results.

The classic experiment is not widely used on the factory floor for the improvement of designs. The time and expense of the procedure has made it unattractive to many companies, especially smaller companies that are typically short of resources. A classic experiment usually requires the isolation of a production process for testing. In addition, the process of investigation involves conducting dozens of separate experiments that are costly to any company. Another problem with the method for classic experiments is that complex statistical models are needed to properly analyze the results requiring people who are skilled and experienced with experimental design to use and interpret the results. Such people are difficult and expensive to locate.

THE TAGUCHI SOLUTION

The shortfalls of classic experimentation for production experimentation have been addressed in a dramatic fashion by Dr. Genichi Taguchi. Dr. Taguchi solved many of the problems of the classical experiment by redefining the assumptions about quality, factors, effects, and design.

Dr. Taguchi sees quality as the "loss experienced by the society after the product is shipped." Dr. Taguchi makes the statement that a company that fails to design products for the maximum benefit of society is worse than a thief. When a thief steals money, there is no gain or loss to the society, because the money has merely changed hands. However, a company that reduces the quality of a product to save $10,000 when the extra repairs for customers will cost $15,000 has imparted a $5,000 loss to the entire society. Dr. Taguchi's comment is an excellent example of how the Japanese value the social unit above the needs of the individual. This belief is reflected in much of Dr. Taguchi's work.

Dr. Taguchi began his new system when he was the head of research for the Japanese Telephone Communications Network. In 1950 he was ordered to improve the state of the Japanese communications system as part of the American Occupation agreements. Upon examining the situation, Dr. Taguchi determined that a 20-year research and development effort would be needed to redesign the communication system using the classic experimentation methods. Therefore, he adopted a faster and more systematic approach to experimentation. The central problem, as he saw it, was working with all combinations of factors and there were simply too many.

To solve this problem, he utilized "orthogonal arrays." Orthogonal arrays are mathematically balanced samples of combinations of factors. This "balance" means that the arrays can select representative groups of factor settings. The precision of a full-scale experiment is compromised for a quick picture of general effects. In other words, some information quickly is better than all information delayed.

Design is seen more as a product planning system. Dr. Taguchi advocated that all designs begin with the product. Reliability and quality engineering of the *product design* are stressed since a good design prevents costly problems with the product. The second stage of the system is to *design the parameters* of the processes that will create the product. By finding the most economical settings for the highest quality, the efficiency of the process is maximized. The third step is to test for the *optimal tolerances* to be used for the characteristics of the product. This step is last because it requires extra cost. If a particular part or component

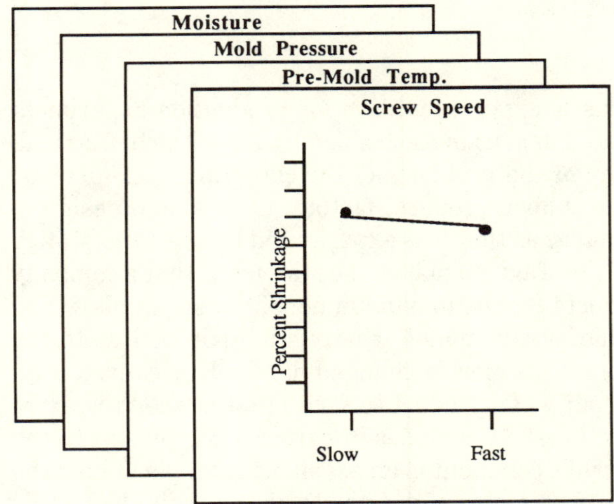

The effect of each factor tested in an experiment can be seen using an effect diagram.
Each level of the factor involved is plotted for its effect on the results being measured.
Above, screw speed has little effect on results.

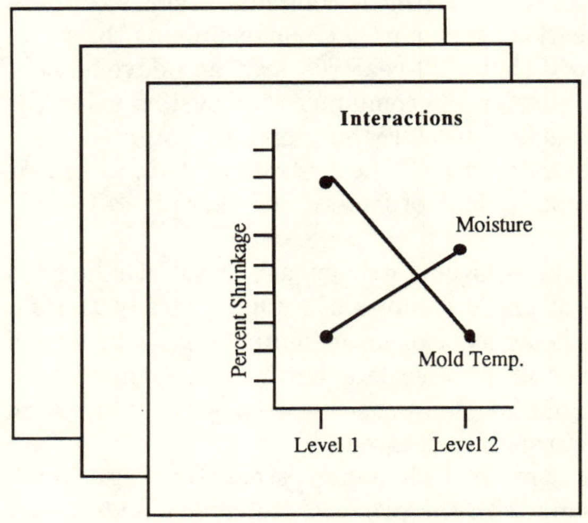

The interaction of two factors can also be plotted. Above, a strong interaction
between pre-molding temperature and moisture is discovered. By studying
such charts and conducting detailed statistical tests, the engineer can find
optimal settings for the process, make improvements in design, and test the
effectiveness of tolerances.

Figure 10.2 Using effect diagrams for examining test results.

In a full factorial experiment every possible combination of factors is tested. Therefore, the number of experimental runs required can be calculated by using permutation equations. For example, a plastics engineer wants to test a plastic molding machine for four factors, each set to either a high or low level. The number of permutations would equal two to the power of four, or 16.

Assuming the four factors are screw speed, pre-mold temperature of the plastic, mold pressure, and moisture, the following experimental design would be created:

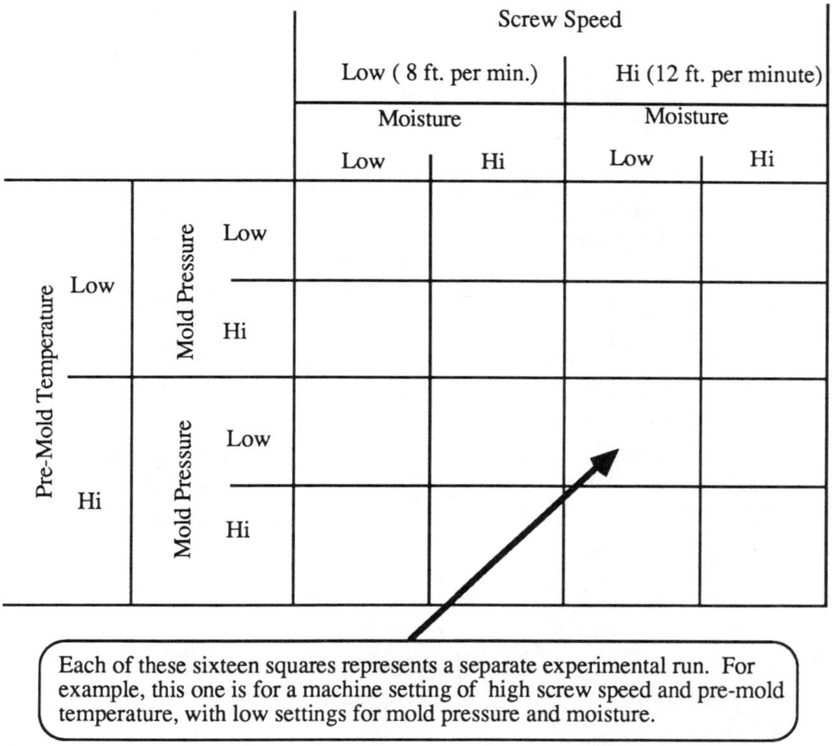

			Screw Speed			
			Low (8 ft. per min.)		Hi (12 ft. per minute)	
			Moisture		Moisture	
			Low	Hi	Low	Hi
Low	Mold Pressure	Low				
		Hi				
Hi	Mold Pressure	Low				
		Hi				

Pre-Mold Temperature

> Each of these sixteen squares represents a separate experimental run. For example, this one is for a machine setting of high screw speed and pre-mold temperature, with low settings for mold pressure and moisture.

In addition to the analysis for each factor, each interaction must be calculated through a complex equation.

Figure 10.3 An example of the design for a full factorial experiment.

is sensitive to variations, extra money is required to ensure only the highest quality materials are used.

This three-step system of design embodies the three main philosophies of Dr. Taguchi. First, the product design should be robust against variations in the process that are unavoidable. These variations could be

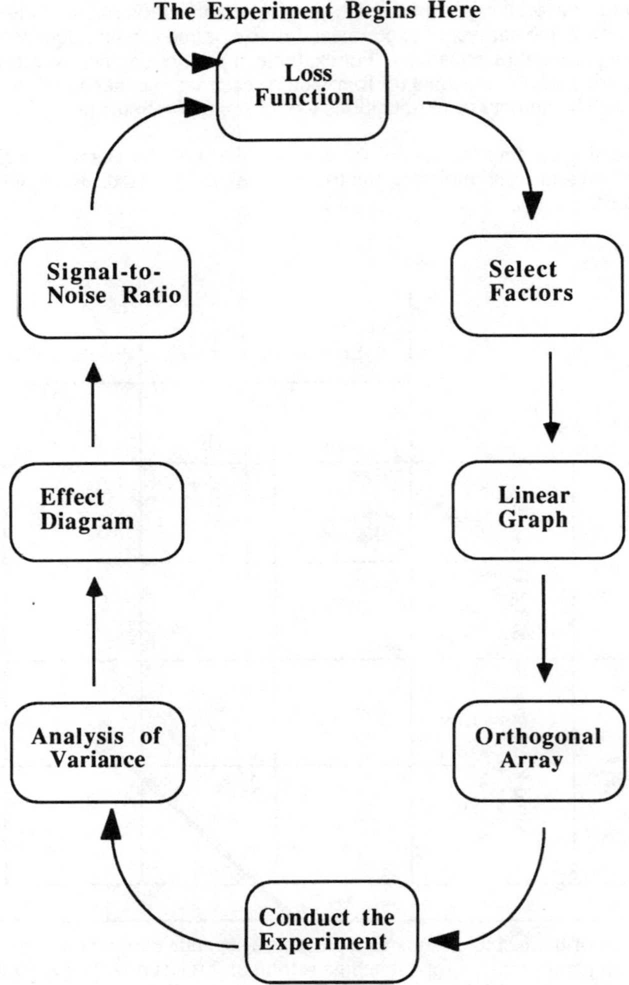

The process of using Taguchi's methods of experimentation can vary widely according to application. Above is a common example of the steps involved in a Taguchi experiment. Computer software can eliminate the need for hand done statistical calculations. In addition, Taguchi's use of orthogonal arrays makes the design of the experiment fast and simple. Thus, the simplicity, flexibility, and practicality of the cost saving approach that is the hallmark of Taguchi.

Figure 10.4 The process of the Taguchi experiment.

factors such as changes in material, operators, and the environment. Dr. Taguchi calls these factors "noise." For example, instead of building air conditioning equipment to protect a large computer from temperature variations, engineers should design the computer to be robust against temperature changes which would eliminate the extra expense of the air conditioner. The second principle is that processes and tolerances should be at the most economical level possible. An experiment may show that increasing the production rate of a machine by 10% does not affect the quality of production. Therefore, it is more economical to the company to run at the higher rate. The third part of the Taguchi philosophy is that experiments should be a regular part of all operations within a company. Taguchi's methods allow people at the technician level to conduct experiments. For a large company, only one highly experienced person is needed to audit the appropriateness of the experiments.

From Capability Study to Loss-Function

Dr. Taguchi's idea of a loss to society is measured using a simple quadratic equation to estimate the economic impact of parts made deviant from an optimal tolerance. The information for the calculation of the loss-function comes from an established SPC system. Dr. Taguchi is very clear on requiring an SPC system to provide two important pieces of information. The first is the current capability of the process or product under study allowing the loss-function to be calculated to evaluate the economic impact of the product or process so that later, the improvements from experimentation can be evaluated directly. Also, this information from past products can be used to make system design decisions for new products. The other source of information is the record of machine settings used to correct problems in the past which is especially critical for selecting characteristics and factors to study during parameter design.

Calculating a Loss-Function

In Figure 10 - 6 there is an illustration of how the loss-function is at odds with the Cpk index. The normal distribution of wires above center is for six inches and is mostly within tolerance limits. A Cpk of 1.33 would be about right and would be rated as a capable process. However, Taguchi sees this as a potentially risky process. The dangerous assumption lies in thinking that the tolerances are magic points—suddenly, parts outside of them are bad, while all parts within the limits are equally good. Taguchi says that this condition cannot be so. Perhaps a wire cut to the low end of the distribution will cause a problem when it is soldered across a bridge that was made to the high end of its distribution. The result is a weaker

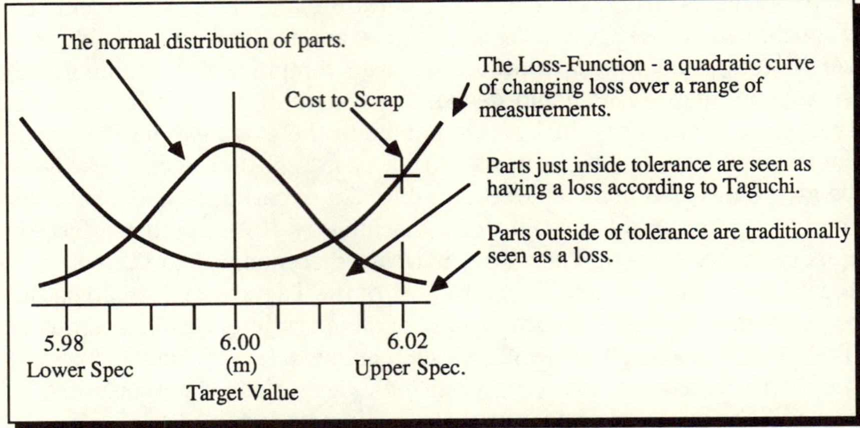

Loss - Function for specific parts: To calculate the loss created by a specific part not being manufactured to the target value, the following formula is used:

$$L(y) = k \, (\, y - m \,)^2$$

where,   L(y) = the loss function
m = the nominal (target) value
y = the size of the part in question

$$k = \frac{A}{\Delta^2}$$

with, A = the cost to scrap a part
Δ = the distance from the nominal value to the tolerance

Loss - Function for total loss for average part : The effect of the process not being on target and the effect of variation can be evaluated using the following formula:

$$L(y) = k \, [\, \sigma^2 + (\, \bar{y} - m \,)^2 \,]$$

where, L(y) = the total loss function

y = the process average
m = the nominal (target) value
σ = the standard deviation of the process

$$k = \frac{A}{\Delta^2}$$

with, A = the cost to scrap a part
Δ = the distance from the nominal value to the tolerance

Figure 10.5 Two of the Loss-Function formulas.

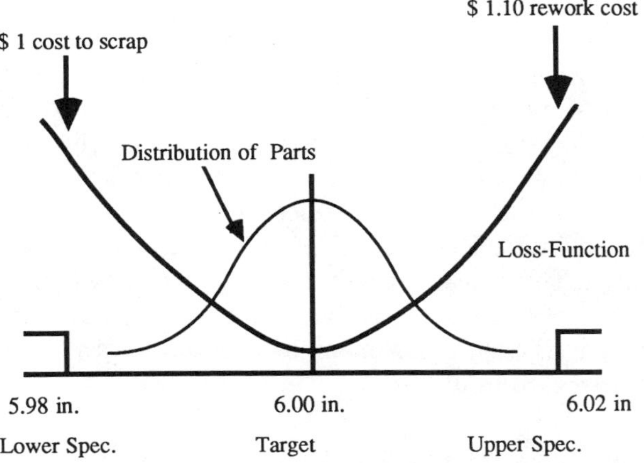

The Loss-Function at Work

The loss-function for a single piece can be calculated
by using the formula:

$$L = k\,(\,y - m\,)^2$$

Where,
  L = the loss-function
  k = a constant, represented by the cost of
      scraping a part divided by the square
      of the tolerance range
  y = the average value during production
  m = the nominal value on the specification

For example, if specifications called for a six inch
piece of tube stock, production cuts a piece to 6.01
inches, the cost of scraping a part was $1.00, and
the tolerance was ±0.02 inch, then the formula
would be:

$$k = 1.00\,/\,0.02^2 = 2500$$

$$\text{and,}\quad L = 2500\,(6.01 - 6.00)^2 = \$0.25$$

Figure 10.6 Example of the calculation of a loss-function for the loss for
a specific piece.

component and possibly a repair bill to the customer (a loss). The loss occurs even though the wire and the bridge were both made within specifications.

The loss function encourages three actions. First, production of parts and components is targeted to cluster tightly around an optimal setting, as opposed to "staying in tolerance." Second, the economic impact of lower quality is directly expressed. Instead of the cryptic Cpk, dollars are used as a measure of quality. Finally, the loss-function makes no assumptions about tolerance limits. Every specification must be tested and evaluated within a design. Tolerances are only created with knowledge. Gone is the former practice of assigning standard tolerances to standard characteristics, such as assigning a plus or minus one-hundredth of an inch to any dimension expressed in inches.

Designing the Experiments: The Orthogonal Array

The loss-function is always the first and last step in any experiment conducted using Taguchi's methods. During the first application, the loss-function summarizes the current situation. Later it will evaluate experimental results. The wire cutting example from Figure 10-7 illustrates the use of the loss-function within the design of the experiment. To begin, the loss-function calculated that the average loss per part under the current conditions of the process centered at six inches and a standard deviation that was one-eighth of the tolerance was $0.01625. This is roughly a penny and a half per piece. This doesn't seem like much until it is taken in the context that the machine makes two and a half million units a year amounting to a total loss of over $40,000 a year.

Alerted to the need to improve the cutting machine, an engineer begins to experiment with the process. Upon examining the process and talking to operators, the engineer states the problem at hand. In this case, variations in the length of wire being cut by the machine are within tolerance, but creating a loss. The objective is to find a way to set up the machine to produce less variation and this begins with the experimenter selecting the factors for study.

The engineer selects three factors for study. Factor "A" is the use of measurement devices by the operators to determine length. The optimal size is stated in the blueprint as "6.00 inches (12.70 mm)". Because of this double statement, some operators use metric calipers to measure; others use English calipers. The experimenter will use both types to see if either makes a difference. Factor "B" deals with the way the cutting blade is set up in the machine. Some operators set the blade with its beveled edge to the right; others set its beveled edge to the left. The machine allows the blade to be mounted either way. Factor "C" is the

A wire cutting machine is supposed to cut wires to a length of six inches. However, a capability study reveals that the machine is actually cutting wires to an average length of 6.001 inches with a standard deviation of 0.005 inches. The tolerance for the wires is set to ± 0.02 inches. It costs $0.25 to scrap or rework a piece. The situation looks like this.

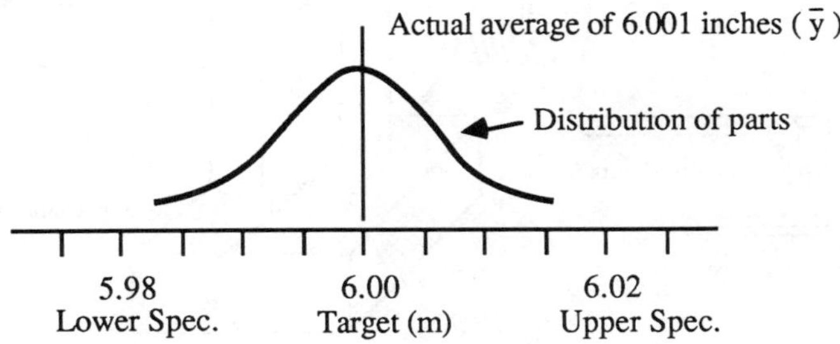

The loss-function for the effect of the process being off target and the variation of the wires would be calculated as,

$$L(y) = k \; [\; \sigma^2 + (\bar{y} - m)^2 \;]$$

$$y = 6.001$$
$$\sigma = 0.005$$
$$m = 6.00$$
$$k = \frac{0.25}{.02^2} = 625$$

thus, $L(y) = 625 \; [\; 0.005^2 + (6.001 - 6.00)^2 \;]$

$$= 625 \; [\; 0.000025 + 0.000001]$$

$$= \$ \, 0.01625$$

in other words, the typical part in this process creates a loss of a little over sixteen cents. With production of 2.5 million wires a year the total loss is about $ 40, 625.

$$\$ \, 0.01625 \; \times \; 2,500,000 \; = \; \$ \, 40,625$$

Figure 10.7 The calcuation of a loss function for the wire cutting experiment. The experimenter now has a rough estimate of the economic impact of process.

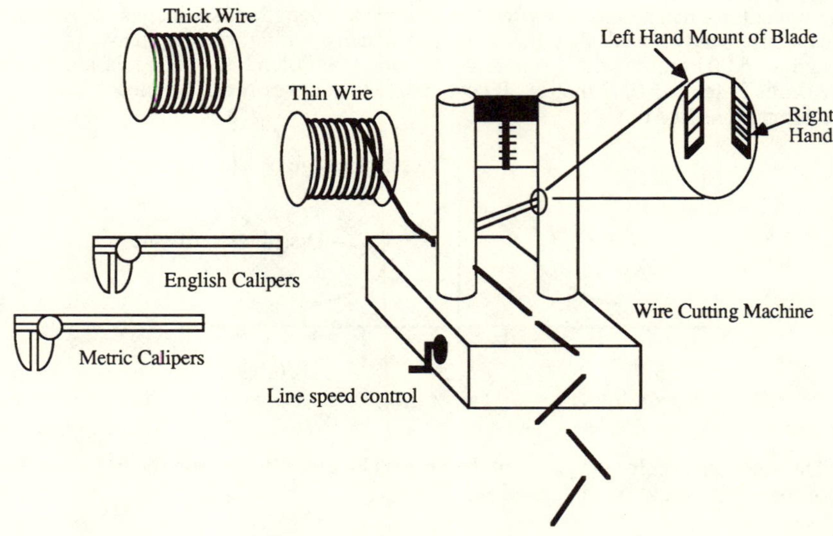

The experimenter now examines the process to select factors for use. In this case the method of measuring (Factor A), the method of mounting the cutting blade (Factor B), and the thickness of the wire being cut (Factor C) are selected for experimentation. The line speed of the machine was examined, but rejected.

Each of the selected factors exists as an either/or condition. Therefore, only two levels of each factor are possible for the experiment. Thus, the factors are divided into the following groups:

Factor A_1 = **The Metric calipers**
Factor A_2 = **The English calipers**

Factor B_1 = **A Right-hand mounting of the blade**
Factor B_2 = **A Left-hand mounting of the blade**

Factor C_1 = **Thin wire**
Factor C_2 = **Thick wire**

Figure 10.8 The selection of experimental factors.

type of wire used. Most of the current wire comes from the Farley Company and has an average diameter of 0.050 inch. A less frequently used wire from the Xeron Furniture Company has an average diameter of 0.060 inch. Part of the experiment will see if wire diameters are important to cutting success.

Other factors were considered and rejected such as when the machine can cut at various speeds, but production requirements state that one speed is to be used to maintain a low inventory of staged wires. Therefore, the experimenter knows that changing machine speed may improve the process, but it will do greater harm to the flow of materials within the plant.

The engineer decides that a two-level experiment is necessary since precision is not important. A two-level experiment means that each factor will be tested at each of two levels. For example, the metric caliper will be level one (A1) and the English caliper level two (A2). The way the blade is set in the machine is either right-handed or left handed. Level one (B1) will be right-handed, level two (B2) will be left handed. Thin wire (0.050'') is selected as level one (C-1) for wire diameter, and thick wire (0.060'') for level two (C2).

The engineer now addresses the problem of possible interactions of these three factors. The engineer's greatest curiosity is about the effect of the way the blade is mounted and the thickness of the wire. This interaction is stated as "B × C" or the "interaction of Factor B and Factor C." To be safe, the engineer also selects "A × B" and "A × C" interactions in case the type of measuring device used proves significant. The following chart illustrates the factors and interactions selected.

The physical design of the experiment is now conducted. The engineer selects an orthogonal array for the study of three factors and three interactions. Using a standard set of orthogonal arrays, the L8 design seems to be best suited for the job. The L8 array will require at least eight experimental runs.

The diagram next to the array in Figure 10-10 is called the "linear graph." Its purpose is to help the engineer place the factors into the appropriate column for creating the design. Each circle on the linear graph represents a possible factor and each line connecting circles represents an interaction. For example, the engineer chooses to place Factor "A" on the circle marked "1." This means that Factor "A" belongs above column one on the orthogonal array. Factor "B" is assigned to circle "2" or the second column in the array. The line connecting "1" and "2" is marked "3." It represents the interaction between the two factors, therefore, column three in the array is marked for the "A × B interaction."

This particular array is filled with "1s" and "2s." The 1s represent the first setting for a factor and the 2s represent the second setting for a factor. Going down to the third row in the array finds the combination of 1,2,2,1,1,2,2. These would be the settings for one experiment. All eight rows are used in the complete experiment. In the example, only columns 1, 2 and 4 represent factors. The design of the experiment is taken from the suggested settings of factors in the row. If the engineer randomly selected this row as the first experimental run, metric calipers would be used (level 1), the blade set left-handed level 2), and thin wire used (level

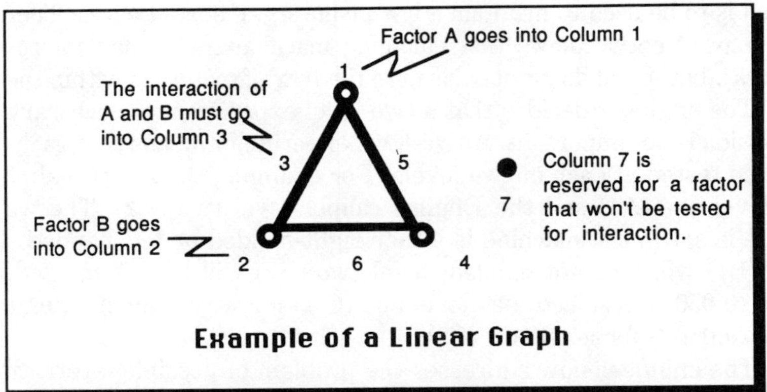

A linear graph helps an experimenter in the design of an experiment. The linear array is a map of the interactions between factors that can be derived from Taguchi's style of experimentation. For example, the circles indicate the possible number of factors that can be studied in an experimental design. The lines connecting circles indicate how many interactions can be included in the design. This linear array above is for an L_8 Array. There are four circles with three connecting lines therefore, up to four factors and three interactions can be studied with this design.

Using the factors and interactions described in Fig. 10 - 8, the following linear graph is created.

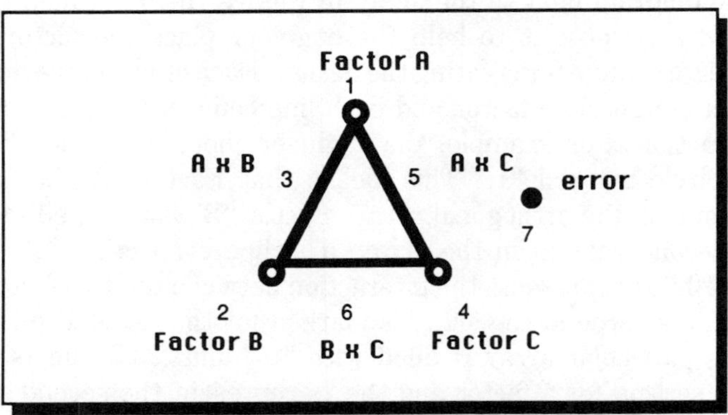

Note on circle seven the term "error." All unassigned points or lines become factors for error.

Figure 10.9 The use of a linear array to begin the design of an experiment.

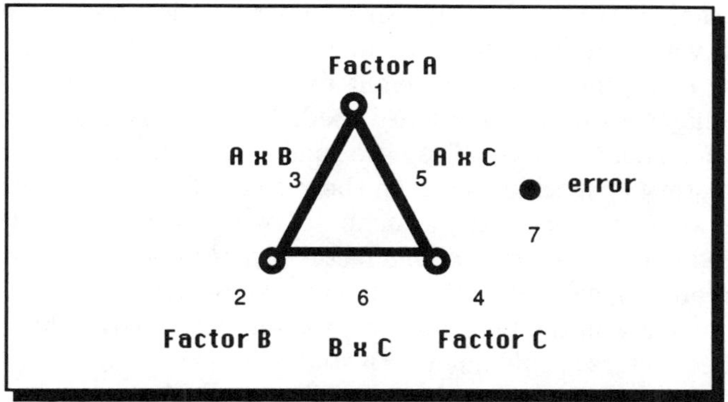

Once the linear graph is completed, it is used to design the actual experiment. Each linear graph has an accompanying orthogonal array. Below is the L_8 array for the linear graph above. Each column on the array corresponds to a point or line on the linear graph. Note how each factor or interaction above is paired with the proper column below.

An L_8 orthogonal array requires that eight experimental runs be conducted. These are the row numbers. Each row has a series of numbers. Those that fall within a factor column determine how that particular experiment should be run.

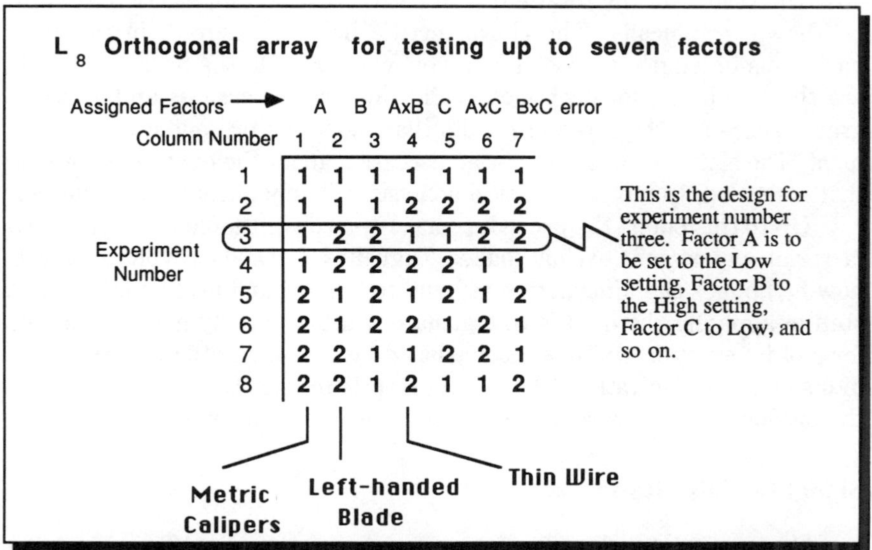

Figure 10.10 Using a linear graph to place factors on the orthogonal array.

1). Using these settings, the engineer would cut some wires and measure their total variation. This information is then recorded in the space following the row being used. For eight experimental runs, the different settings suggested by the array are used. In this example, each run is repeated in a random order. They are randomly selected to prevent the effect of history or machine wear on the results. The last column of this particular array is filled with a small "e" which represents the error experienced because every possible factor has not been listed. In all arrays, any leftover columns are assumed to represent this error. Should the error be large in the final analysis, the engineer knows that some of the excluded factors should have been studied. To study all of the factors, a full factorial experiment would be required.

Analysis of results is conducted by separating the effect at level 1 from the effect at level 2. Figure 10-11 shows how the array marks are grouped together by results. The result total of level 2 is subtracted from the result total from level 1. The difference is then squared. This value is divided by the number of experimental runs. This procedure calculates the *sum of squares*. The sum of squares allows the experimenter to use an Analysis of Variance (ANOVA) table. The ANOVA table calculates the total effect for each factor, interaction and error. (see example below).

The results of the total effect of each factor and interaction can also be shown graphically. The characteristic being measured, in this case variations in length, is placed on a vertical scale. On the horizontal scale are the two levels for each factor. Plotting results creates an effect diagram. These are the types of results that should be presented to management. The simple charts and a brief discussion about the experiment are all that is needed. Complex statistical analysis will only alienate an audience.

Up to this point, the procedure has been conventional for a fractional factorial experiment. What makes Taguchi's methods revolutionary is how he applies the information at hand to the original problem. The first step in applying the data is to translate it into a clearly understandable form of information which is accomplished by converting the ANOVA data into signal-to-noise ratios. The second step is to use the results to change the product or process specifications to their most economical settings.

Signal-to-Noise Ratio

"Noise" is defined by Taguchi as "variables that disturb the function of a product." All the factors that are beyond the normal control of the company, such as variations in the weather, materials, operators, or raw products are included. The error columns are used in the calculation of this background "noise". The "signal" is considered the strength of a factor

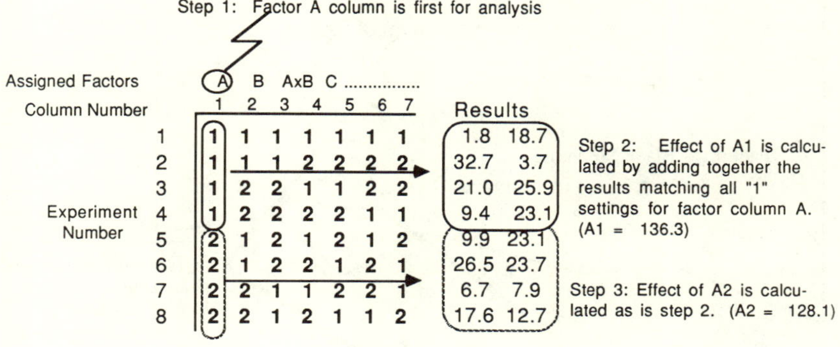

Step 4:   Calculation of the Sum of Squares for factor A.
This is done by squaring the difference between A1 and A2 and dividing the results by the number of experiments.

$$SS\,a = \frac{(A1 - A2)^2}{16} = \frac{(136.3 - 128.1)^2}{16} = 4.2025$$

The results of each experimental run are placed after each matching row. These are used to calculate the sum of squares for each factor. The methods of determining the sum of squares will vary across designs. Above is an example of calculating for Factor A; below is the calculation for Factor B. Other factors and interactions are calculated for later analysis by ANOVA or Effect Diagrams.

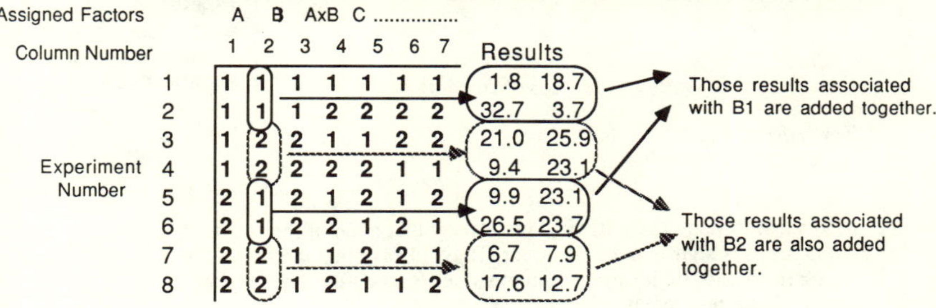

The effect of factor B is calculated using the Sum of Squares.

$$SSb = \frac{(B1 - B2)^2}{16} = \frac{(140.1 - 124.3)^2}{16} = 15.60$$

Figure 10.11 Calculating results on an orthogonal array.

being tested. The ratio of the strength of the signal divided by the power of the error calculates the signal-to-noise ratio and is the same signal-to-noise ratio used with communication equipment or the home stereo system. The higher the ratio, the more efficiently the signal overwhelms the noise. Taguchi even translates these signal-to-noise ratios into decibels. Those factors displaying high decibels will control a product or process best.

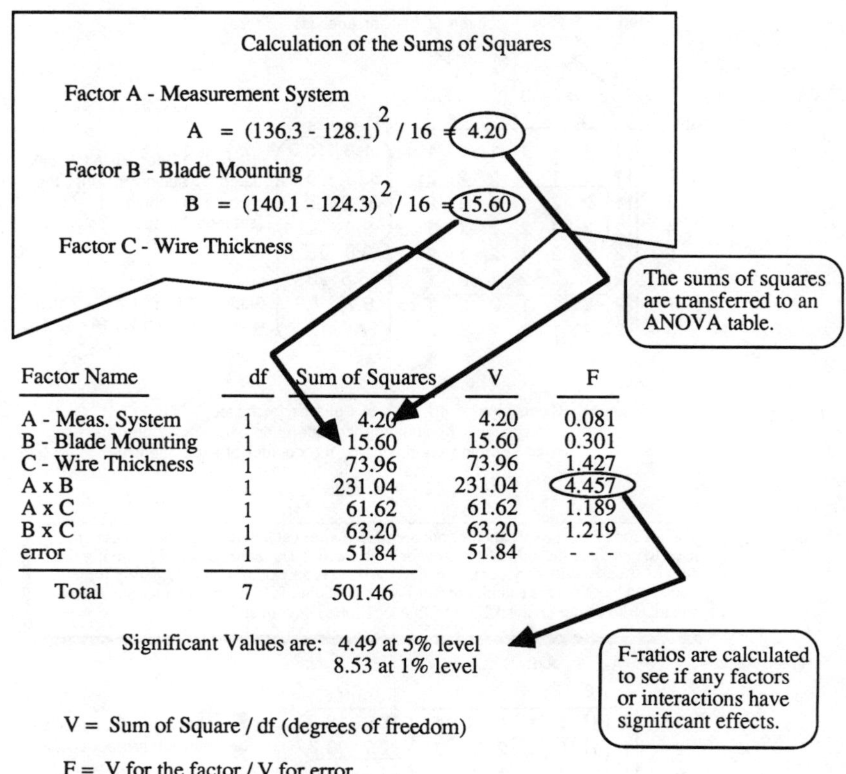

Factor Name	df	Sum of Squares	V	F
A - Meas. System	1	4.20	4.20	0.081
B - Blade Mounting	1	15.60	15.60	0.301
C - Wire Thickness	1	73.96	73.96	1.427
A x B	1	231.04	231.04	4.457
A x C	1	61.62	61.62	1.189
B x C	1	63.20	63.20	1.219
error	1	51.84	51.84	- - -
Total	7	501.46		

Significant Values are: 4.49 at 5% level
8.53 at 1% level

V = Sum of Square / df (degrees of freedom)

F = V for the factor / V for error

In this experiment, the statistical results say that none of the factors or interactions is significant. However, Taguchi says that the factors or interactions with high sum of square values (especially those higher than error) should be seen as affecting the process.

There are several steps that could follow. In Fig. 10 - 13, effect diagrams are used to determine the optimal settings for this process. In Fig. 10 - 14, signal-to-noise ratios are used for the same purpose. Another possibility would be to pool together low effect factors above and recalculate ANOVA.

Figure 10.12 The statistical analysis of test data using Analysis of Variance (ANOVA).

In the results, each factor either has a good signal-to-noise ratio or it doesn't. Both results are important for making the most economical setting possible. The previous wire cutting example can illustrate how this is done. The effect of using either metric or English measurement systems was slight. Therefore, the engineer chooses the system that is

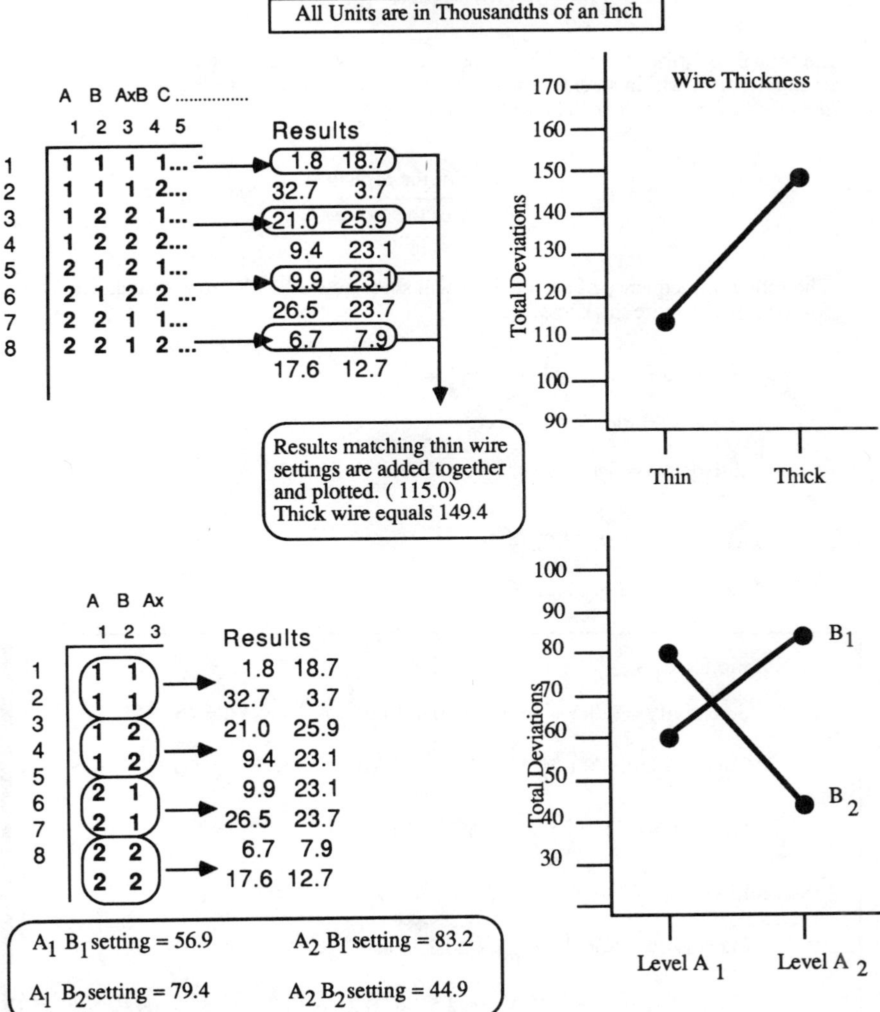

The effect of each factor on total variation can be shown in the above effect diagrams. The lower the total deviations the better. Since there is strong interaction between factors A and B, the lowest combination is chosen. English and the left-handed mounting of the blade reduce variation so, it also is choosen. Thin wire greatly reduces variation so it too, is selected. Thus, the optimal settings for this process are thin wire, left-hand blade, and English calipers.

Figure 10.13 Seeing results with effect diagrams.

The Signal-to-Noise Ratio

The Signal-to-Noise Ratio (S/N) is a measure of how strong a factor is over the amount of error within an experiment. This error is seen as the background "noise" of the experiment. The formula for S/N is,

$$\text{S/N Ratio} = \frac{\text{Power of the Signal}}{\text{Power of the Noise}}$$

The following sequence of calculations will show how a S/N Ratio is calculated for the effect of wire thickness.

$$\text{Sensitivity} = (x_1 + x_2 + x_3 + \ldots + x_n)^2 \, / \, n$$
$$\text{Total Variation} = (x_1^2 + x_2^2 + x_3^2 \ldots + x_n^2) - \text{Sensitivity}$$
$$\text{S/N Ratio} = \text{Sensitivity} / (\text{Total Variation} / (n - 1))$$

C_1 data = 20.5, 46.9, 33.0, 14.6

C_2 data = 36.4, 32.5, 50.2, 30.3

S/N Ratio for C_1 -

$$\text{Sensitivity} = (20.5 + 46.9 + 33.0 + 14.6)^2 \, / \, 4 \; = 3306.25$$

$$\text{Total Variation} = (20.5^2 + 46.9^2 + 33.0^2 + 14.6^2) - 3306.25 \; = 615.77$$

$$\text{S/N Ratio} = 3306.25 \, / \, (615.77 / 3) = \boxed{16.11}$$

S/N Ratio for C_2 -

$$\text{Sensitivity} = (36.4 + 32.5 + 50.2 + 30.3)^2 / 4 \; = 5580.09$$

$$\text{Total Variation} = (36.4^2 + 32.5^2 + 50.2^2 + 30.3^2) - 5580.09 \; = 239.25$$

$$\text{S/N Ratio} = 5580.09 \, / \, (239.25 / 3) = 69.97$$

Since lower variation is the goal of this experiment, the lower signal-to-noise ratio is selected. In this case, the 16.11 of level 1 (thin wire).

Figure 10.14 The calculation of signal-to-noise ratios.

cheapest to use. Since all the operators are familiar with the English system, and most of the calipers in the shop are English, the English system is selected. No extra money is required to train personnel in the use of metric, or to buy more metric gages. The left-hand setting for the blade reduced variations significantly, so this is now specified as, "standard procedure." It costs the same to mount the blade either way, but left-handed reduces variations and loss. Finally, the thin wire worked best to reduce variation. Therefore, only thin wire was allowed for this machine during production.

The engineer would conclude the experiment by again accessing the machine's capability. A reduction in variations in wire lengths would be confirmed and a new loss-function calculated. The difference between this loss and the one estimated before experimentation would show the total impact of the experiment on cost reductions.

Further Observations

The above discussion only begins to reveal the full potential of Taguchi's methods. The illustrations have deliberately focused on the simplest form of experimentation where only two levels of each factor are tested. There are orthogonal arrays that can evaluate dozens of factors and interactions, each at several levels. Such experiments would explore the effect of factors over a range of settings. Most of these responses would be nonlinear so that even more options would be available for selecting proper settings or specifications.

The techniques that Dr. Taguchi advocates are relatively simple in comparison to their alternatives. The most difficult thing to explain about Taguchi's methods is the endless potential they have introduced. His methods redefine the role of research and development in any company. Research can proceed at a faster pace, covering more topics. In addition, engineering becomes an occupation for all industrial departments. Tests and studies are encouraged at all phases of manufacturing, but, Taguchi recommends experimentation as early in the process as possible. After all, the main theme of Taguchi's methods is to prevent a loss to society, including an added expense for the manufacturer.

OBJECTIONS TO TAGUCHI

In the United States a considerable controversy is currently raging about the use of Taguchi's methods. The first source of resistance concerns the statistical purity of the methods. Specifically, some statisticians

where, L(y) = the total loss-function

$\overline{y}$ = the process average
m = the nominal (target) value
σ = the standard deviation of the process

$$k = \frac{A}{\Delta^2}$$ with, A = the cost to scrap a part
Δ = the distance from the nominal
value to the tolerance

The experiment has concluded, but the final results remain to be calculated. Specifically, the experimenter must now recalculate the loss-function for this process to find out how much improvement has been realized by the new optimal settings.

A new capability study finds that the process average is still 6.001 inches, but the standard deviation has fallen to 0.004 inches. Assuming the same specification as in Fig. 10 - 7, the new loss-function is,

$$L(y) = k [\sigma^2 + (\overline{y} - m)^2]$$

$$= \frac{0.25}{0.02^2} [0.004^2 + (6.001 - 6.000)^2]$$

$$= 625 \times 0.000017$$

$$= \$ 0.010625$$

or, total loss is about a penny per piece. However multiplied by one year's production, the loss becomes,

$$\$ 0.010625 \times 2,500,000 = \$ 26,562.50$$

This is a substantial reduction from the loss of \$ 40,625 from the process before experimentation took place.

Figure 10.15 The calculation of the loss function to evaluate the success of the experiment.

realize that orthogonal arrays can easily overlook particular effects in exchange for a general effect. The resulting predictions of effect can then be confounded or biased. Taguchi supporters counter that the method has demonstrated high repeatability in use. The Taguchi methods do have the advantage that changes in process or products will produce immediate

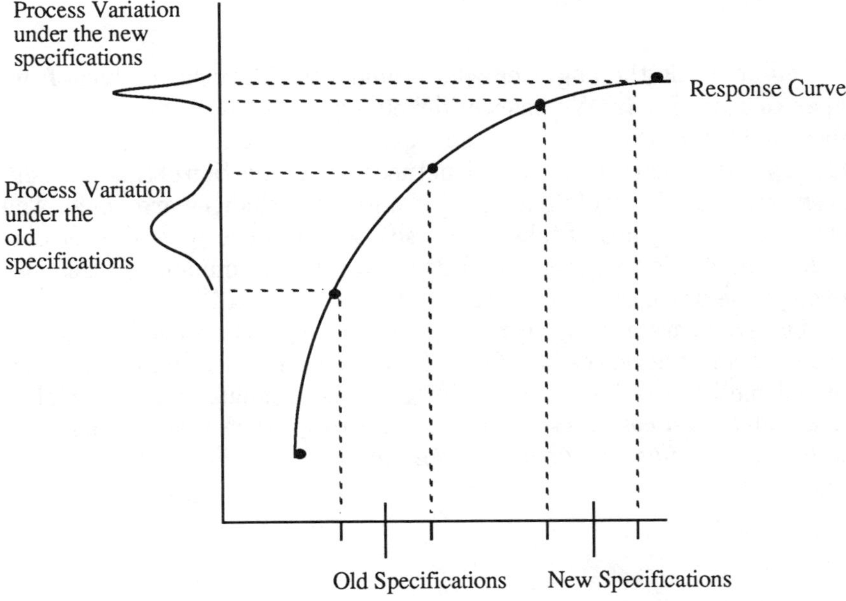

Some experiments will test factors at several levels looking for a non-linear response. When these are found, specifications can be changed to take advantage of this situation. Above, an old tolerance range is moved to the right so that the amount of process variation can be reduced.

Figure 10.16 The use of nonlinear response curves to improve process and product specifications.

effects in production and any mistakes or inaccuracies that are detected can be corrected.

A second complaint about Taguchi's methods is that they are difficult to learn and use. A typical training seminar runs around 80 hours. However, part of the problem comes from the skills possessed by American engineers. The American engineer typically receives little training in experimental design or advanced statistics as part of an engineering degree. These skills are vital for using Taguchi's methods effectively. For a person familiar with experimental design, Taguchi's methods border on eloquent. The idea of directly translating results into meaningful action converts experimentation into a more practical exercise.

SUMMARY

Yu-in Wu is the engineer who brought the Taguchi methods from Japan to Taiwan. Mr. Wu has worked closely with Dr. Taguchi for many years and offers this insight, "The Japanese do nothing by emotion, always by calculation." The Taguchi methods are a reflection of this attitude. Every situation is carefully evaluated and then, changes are made. The concerns of the company take a back seat to the concerns of the society. Lack of quality is regarded as a loss to society and minimizing this loss is the prime objective of Taguchi's methods.

Dr. Taguchi's methods can enable a company to improve its designs, processes and tolerances at a faster rate than other research and development methods. After all, good design is the ultimate source of higher quality and lower costs. As writer John McElroy observes about Taguchi's methods, ". . . vast opportunities for improvement lie beyond the obvious . . ."

CHAPTER 11
MANAGEMENT AND SPC

Implementing SPC is made more difficult by some of the characteristics of today's business management styles. As stated before, top management's commitment to the SPC system is vital for its success. However, one of the rewards of obtaining a top management position is that it insulates the executive from the work force. Middle management's unspoken goal is to keep the problems of employees and productions from upsetting top managers. Consequently, top managers can easily find themselves planning new directions that are far out of step with the potentials and desires of the company employees.

Another inhibitor of SPC comes from top management's ability to hire middle managers. Commonly, top managers tend to avoid hiring someone who will disagree with their views; they tend to hire people who are similar to themselves. The lack of variety in thinking styles will choke the innovation and creativity that is vital to finding the path to continuous improvement. A well-developed SPC system can overcome these problems when it taps a very resourceful and diverse pool of new ideas—specifically, when the SPC system forces top management to listen to the work force.

MANAGEMENT BY WANDERING AROUND (MBWA)

Tom Peters and Robert Waterman in their book *In Search of Excellence* noted that a well-run company was marked by top executives who wandered around. These managers were not comfortable being in an office all day. Instead, they went out and met with customers and employees to learn their needs. In fact, this technique and others from their book are now being widely studied by managers around the world.

MBWA is a very supportive technique for an SPC system. Top management is responsible for planning and resource allocation for the SPC system. By talking to the people on the shop floor and reading the back of the control charts, managers can learn a great deal about the realities of making processes capable of exceeding specifications. When upper level managers do not wander around the shop other problems can result—

in particular, the lack of clear communication and full understanding of their plans. A true story will illustrate this point. A company president was convinced by the author to visit his third shift at midnight to ask employees if they had questions about their new SPC system. One night he came in the back door dressed in the traditional three-piece business suit. During the 1 a.m. training session he stood up and explained how SPC was helping the company to win more customers and make their jobs easier. Then he asked for questions. One worker in the back stood up and asked, "I like the charts, but who are you?"

PARTICIPATORY MANAGEMENT

SPC is a good introduction to the ideas of participatory management, which is the method whereby management plans and communicates strategic goals of the company to the remaining employees who are asked to participate in making decisions on how those goals can be achieved. SPC is a similar model. The strategic goals of SPC include higher quality, lower operating costs, and increased markets. These are so called "motherhood policies" because few would disagree with their merits. The difficulty in the management of SPC is turning these goals into reality. An SPC system does so by forming work teams to actively seek solutions to problems and to promote the achievement of the stated goals.

Common Vision

SPC is a form of participatory management. Companies that wish to switch to a more participatory form of management will find SPC good practice. Top managers who want to convert successfully should remember that any change in an established form of management will be a major cause of resistance from company staff that will result from removal of a well-established and understood method of management. The familiarity of the old system promoted trust and cooperation. Fear via misunderstanding is reduced when a clearly stated vision of the company's goals with SPC is presented to the employees. This "common vision" helps to restore the trust and cooperative spirit of employees by providing security through a shared belief.

Some of these messages are already being used in many companies. "Work smarter, not harder," "Do it right the first time," or "We take pride in our customers," are messages that move companies in a single, well-understood direction. "Do it right the first time," implies that the effort of all employees is to face problems, find out why they happen and prevent them in the future. In other words, it is just another form of continuous improvement.

However, top managers who are not sincere about their company goals or who fail to support goals with actions are a major threat to success in SPC implementation. A company claims to be committed to zero defects, but ships out defective products whenever the delivery schedule is tight, is sending a clear message to the employees that goal is really "make the schedule."

CONTINUOUS IMPROVEMENT

An isolated and unsupported SPC system can actually cause economic harm to the host company. An SPC system is just another layer of procedure to use during production. There is no time for the real benefits of SPC to be realized. The proper way to implement SPC is to make it the sole replacement for current methods of operations by creating a single theme of manufacturing. The force that makes this possible is the philosophy of continuous improvement. Continuous improvement provides the common vision necessary for an effective change in management styles.

Continuous improvement is an attitude adopted by management and employees and is the idea that nothing is static. Every process has an endless potential for improvement. Every person is responsible for finding these opportunities for improvement and exploiting them. Naturally, there will be obstacles to overcome so the implementation of continuous improvement also involves the removal of inhibitors.

Reasons for Continuous Improvement

Dr. Deming and others have pointed out that manufacturing is in a new economic age. Companies that survive are ones that continually seek and capture new markets. To last a company has to produce lower cost goods of higher quality than its competitors. The best way to accomplish this goal is to seek better ways to operate the manufacturing processes. An increase in efficiency means lower operating costs and enables a company to capture more markets for its products. With larger markets the company can hire more workers and make more money. The companies that try to maintain the status quo find their markets whittled away by competitors. Continuous improvement at least makes a company a moving target.

Both General Motors and Ford Motor Company have adopted the

philosophy of continuous improvement as their corporate operating policies. Both of these companies have backed up their words with action. Unlike years before, the idea of continuous improvement is not being delegated to posters and slogans on coffee mugs. Instead, the policies and procedures developed at these two companies are being redesigned to fit the concept of continuous improvement. For example, Ford's supplier quality assurance program, Q–101, requires vendors to adopt a policy of continuous improvement and to demonstrate its applications within their operations.

Putting the Idea of Continuous Improvement to Work

If one key impact of SPC on management had to be identified, it would be the decentralization of authority. The principles of SPC imply that the management of production processes works best when operators share responsibility for daily operations. Once a policy of continuous improvement is in place at a host company, action must be taken to ensure its use. The most common tool for applying the idea of continuous improvement is depending on work teams and problem-solving methods. Work teams are small groups of employees (usually five to nine people) who are assigned an area of production to study. Their task is to search for problems in that area and to effect solutions. To structure this effort, traditional problem-solving methods are employed.

The most important of the problem-solving methods is management support. The team requires time to meet, equipment to help them, and a channel of communication to other departments. Management must provide this support. An example can illustrate these points. Bob is the supervisor of seven welders in the assembly area of the plant. His manager has told him to form his seven people into a work team whose job is to explore ways to decrease the number of welding defects experienced in the final assemblies. Top management is strongly committed to using problem-solving work teams and shows this commitment by having all members of Bob's team trained in traditional problem-solving methods and providing them with a quiet room to meet in once a week. Bob's manager will join the team at every meeting to help relay requests from the team to other managers.

THE PROBLEM-SOLVING CYCLE

The decentralization of authority through the use of SPC and problem-solving teams implies that the production people have been trained in decision-making techniques. Such training must precede the implementation of work teams. The job of problem-solving is much easier when all teams are using similar tools. Furthermore, managers must be active promoters and participants of the work teams. Without this continuous emphasis on team work, people tend to fall back to the more familiar habit of following the instructions of supervisors. Problem-solving with work teams is very different from the concept of centralized authority. It promotes the sharing of ideas, decisions, and techniques. Individuals are encouraged to question former decisions and occasionally to contradict their supervisors. This approach is such a change from traditional beliefs about management that a long period of careful implementation and adjustment is often necessary.

The technique of problem-solving involves a cycle of activities. Texts on the operation of quality circles or other problem-solving groups can explore this cycle more fully. Below is a brief overview of some of the possible methods used in the problem-solving cycle. The SPC system is an invaluable tool for gathering information needed to drive a problem-solving work team. Management's role is to support and direct this cycle of problem-solving.

Step 1. Problem Identification. The team meets to determine which problems are of concern for the group and to more clearly define the situation and its related problems. The situation can be defined by means of a flow chart of the process in question. A flow chart is an illustration of the steps of the production process under study. Attached to the flow chart should be a brief summary of the people and costs involved at each stage of production. Compiling a flow chart helps the team members get a better feel for the situation and to practice working as a team. Later, this flow chart will help people from outside the team quickly learn the situation.

Bob's work team has compiled the following flow chart and the team wants to further define the problem at hand. Management is concerned about welding defects, but which defects? How many defects are experienced in final assembily? To answer these questions, the team requests the scrap and defect reports from the final assembly area. With this information Bob's team can perform Pareto Analysis.

Pareto Analysis is a sorting process based on the concept of the "trivial many versus the vital few." In other words, of all of the hundreds of possible problems with welds, only a vital few of them will occur in large numbers. Sometimes the frequency of a problem is not the only consideration. Some defects have a larger impact on production costs than others. A small leak in a radiator may happen frequently, but it can be

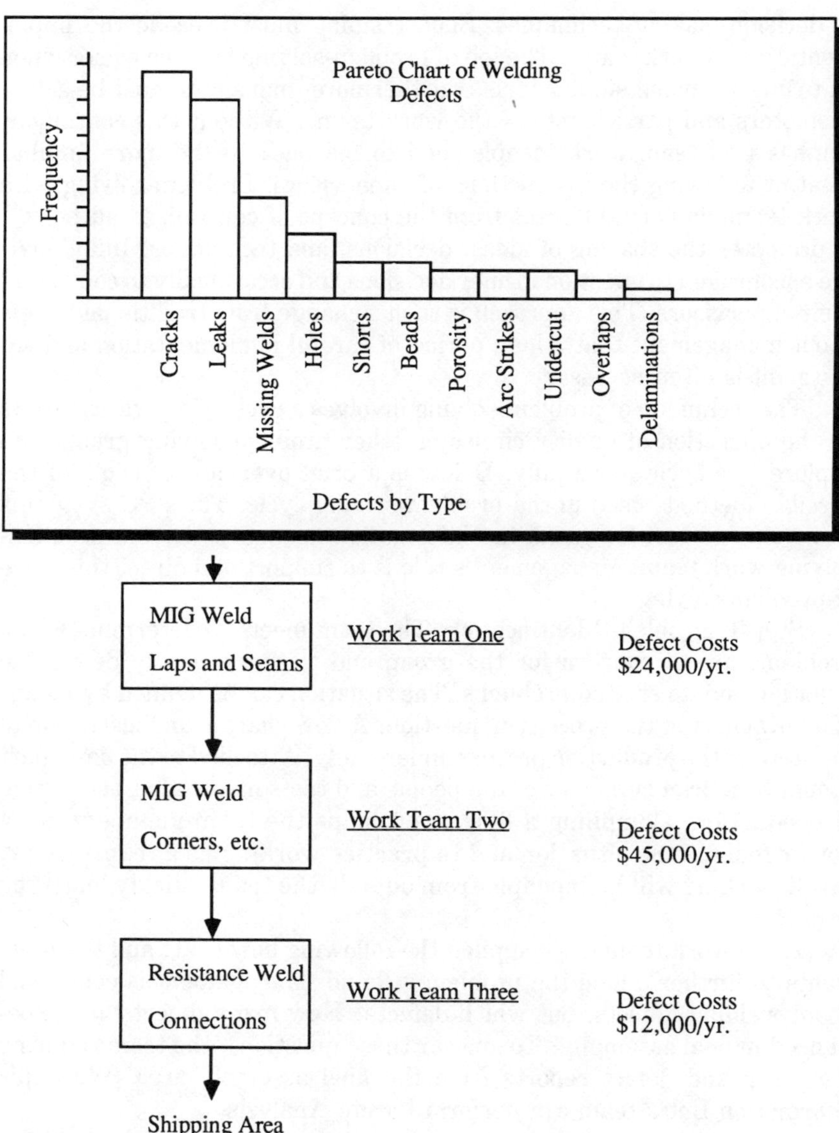

Figure 11.1 Bob's Pareto and flowchart for the welding assembly area.

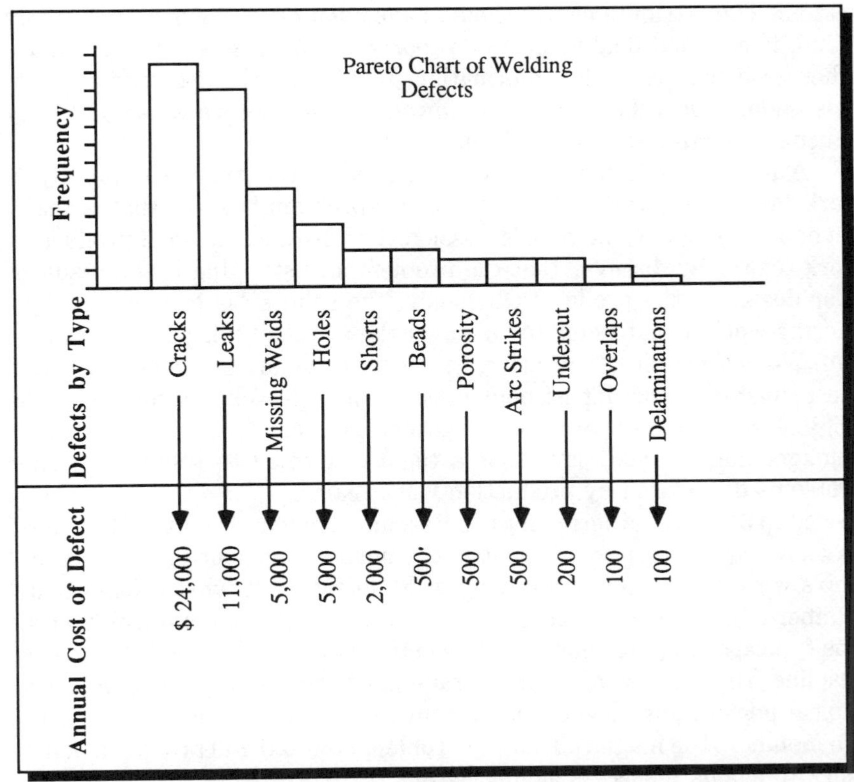

By combining the information for the frequency and cost of each defect, it becomes very easy to find the first defect to attack.

Figure 11.2 Pareto chart with cost estimates.

quickly braised shut for very little cost, but a crack in a radiator that occurs less often, costs substantially more to fix.

Step 2. Problem Selection. Bob's team may find that it needs to select which welding defect to work on first. A small work team with only limited time to operate cannot successfully solve more than one problem at a time, so may use Pareto Analysis to pick the first problem on which to work. One way to do this is to multiply the frequency of each defect by the cost to fix the defect which results in the estimated cost of fixing each defect. The defect that costs the company the most money to fix is chosen first. The other defects are listed in order of expense for action at a later time.

Information for Pareto Analysis is readily available from a good SPC system. The attribute charts should have a listing of the most commonly found defects and final inspection reports and process scrap reports are other good sources of this information. As always, the operator in a process under study should be interviewed, to see if process logs note any unusual occurrences in production.

A note of caution should be mentioned at this time. Frequently, a work team will quickly find an important problem to solve that is really beyond the scope of the team's mission. For instance, a small production work team may discover that the problem it is studying is the result of poor design of the product. Obviously, the solution is to have an engineering work team formed to take over the attack of this particular problem. When these situations arise, the team should arrange a meeting with the manager of the department involved and present its findings. The original work team then selects another problem to study. It is up to management to make sure that a team is formed to solve the larger problems discovered by production work teams.

Step 3. Data Gathering for the Baseline. An important question must be answered at this point; what is the normal behavior of the process? Bob's work team may be studying weld defects, but what is the normal number of defects experienced day after day in the final assembly area? The typical or average number of weld defects found is called the process baseline—in other words, the average performance of the process. This typical performance level is important for the team because it will tell the members the magnitude of the problem now and will provide a benchmark to evaluate success against later.

The x-bar/R charts and attribute charts are a common source of baseline information. If Bob's team decides that cracked welds are the problem to be studied, a c-bar chart in the final assembly area could track the number and types of weld defects. Any action taken by Bob's work team could be evaluated against this chart. The number of cracked welds before and after each action could be compared.

Step 4. Problem Analysis. After the problem is selected and the monitoring charts are in place, a work team must analyze the problem further to draw up a list of possible actions to take. One method is called the "cause and effect chart." Developed by Dr. Ishikawa of Japan, the chart is based on the concept that there are no accidents causing quality problems in a process. A properly designed process will deliberately create high quality at low costs. A problem in a process is the result of a flaw in design.

The chart examines the effect people, machines, management, measurement, methods, materials and the environment have on the process. When constructing the cause and effect chart, the work team members

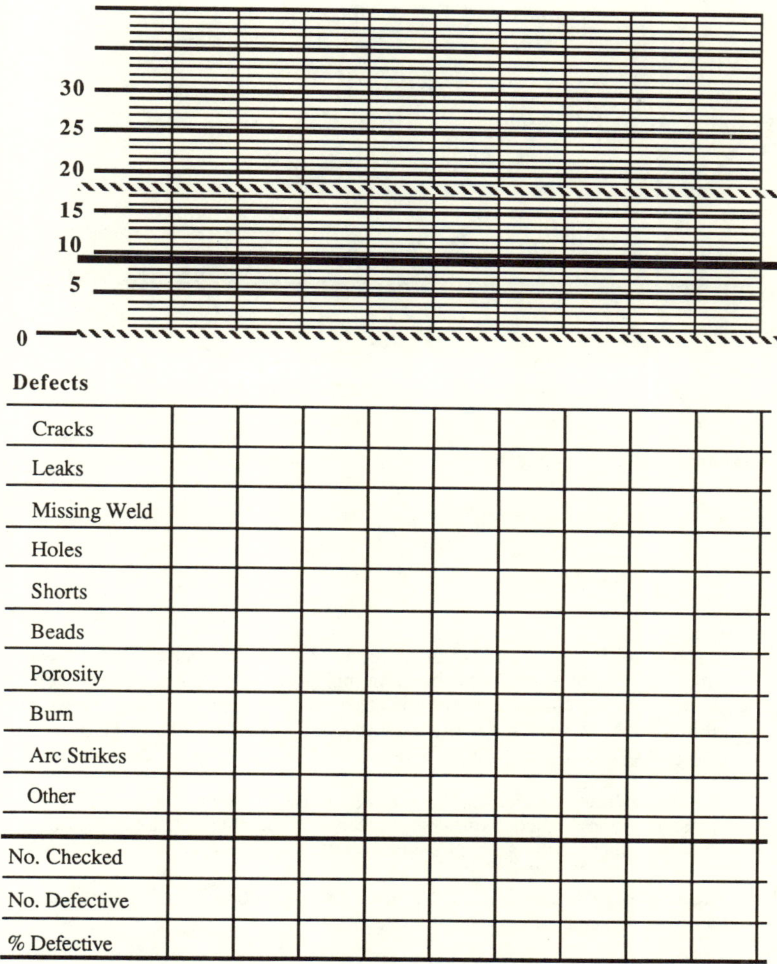

Defects									
Cracks									
Leaks									
Missing Weld									
Holes									
Shorts									
Beads									
Porosity									
Burn									
Arc Strikes									
Other									
No. Checked									
No. Defective									
% Defective									

By studying the number of defects detected in samples of five assemblies, Bob's problem-solving team discovers that an average of nine defects are present. This figure is used to calculate control limits for their c-bar chart.

Figure 11.3 The creation of a baseline from the process average of a c-bar chart.

are asked in turn to think of any possible causes for the problem under discussion. Each idea is placed on the chart where the team feels it belongs. Once completed, the team discusses which factors are most likely the cause of the problem. Bob's team may discuss the effect quotas or inferior welding supplies may have on weld cracks.

Cause and Effect Diagram

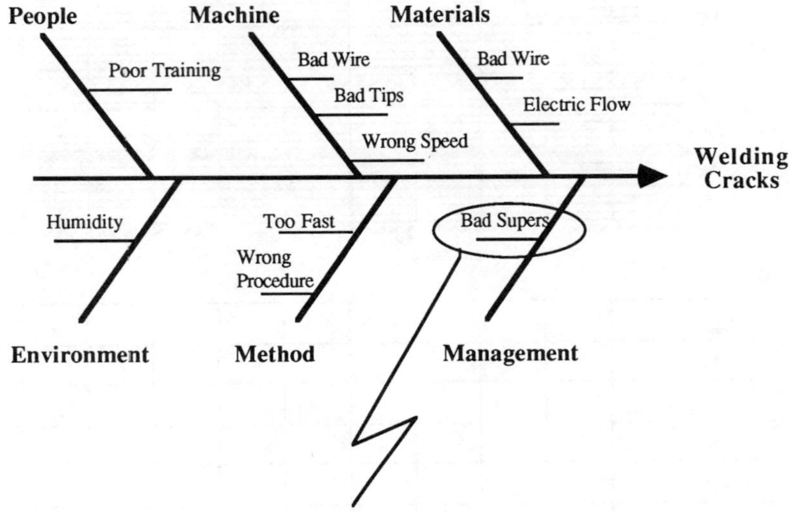

To fill out a cause and effect diagram, the team members must list all possible causes of the problem, in this case, weld defects. Then each idea is assigned to a "bone" on the chart. In this example, one person felt that bad supervision was the cause of the problem. Therefore, it is assigned to the Management "bone."

Also note how one idea can be assigned to two different "bones." The reason they are called "bones" is that the finished charts looks like a fish skeleton. Hence the name, fishbone chart.

Each idea is discussed and voted upon. The resulting list of favorites is used to create an action list for the team.

Figure 11.4 Cause and effect chart.

The discussion of important causes of the problem should lead to a list of possible actions. Each action should be fully described and should include the method of evaluating its success. The team then votes on which action to try first, second, third, and so on creating an *action list* that is a schedule of corrective actions a work team plans to use.

Step 5. Management Presentation. Before the team takes actions it is wise to seek the blessings of management. There are several good reasons for presenting the action list to management. First, it helps keep management informed of the team's progress. Second, a regular series of meetings will keep the problem-solving teams active and motivated. Third, the presentation will help prevent management from accidentally ruining the team's work. A work team could be studying a problem with a product only to learn later that the product is scheduled to be phased out. Finally, a management presentation is a good political formality to help foster a feeling of cooperation between the team and management. A manager hostile towards a work team can be a tremendous inhibiter to the team's success.

Although the entire team should attend, only one member should make the presentation, which should be short and to the point. Illustrations are helpful. The team's main goal in a presentation is to answer a few mandatory questions. Why was this problem chosen for study? Why is the problem important to the company? What is the current level of the problem? What actions are planned by the team? How will each action's success be evaluated?

Once the work team has management's blessings, it can take action.

Step 6. Action. Using the first item on the action list, the work team begins. Bob's team may have voted to examine the effect of the welding wire on weld quality and has already designated a c-bar chart in the final assembly area as the baseline. However, the team has also asked the final assembly people to begin mapping the location of weld cracks on the assemblies. The defect map showed that the cracks occurred in no particular pattern so the team decided that a general cause was at work and the weld wire was one of the favorite suspects.

Bob met with the purchasing agent for welding supplies and discovered that a dozen different types of welding wire were purchased from at least as many suppliers. Bob and the purchasing agent agreed that only one type of wire from one supplier would be purchased and used in his department for a week. The results were immediately obvious on the c-bar chart in the final assembly area. The number of weld cracks dropped to almost zero. The team had found one part of the solution to the weld defect problem.

The team also made another discovery—the wire used was more expensive than most of the other types so that the cost of welding increased. Had the team more than offset the added expense with the reduction in weld cracks? This is an important question for the team to address. Luckily, a good SPC system includes a quality cost report. With such a report, the team can estimate the cost in rework for fixing a single weld crack. Multiplying this figure by the former average number of cracks will provide an estimate of the costs before the new weld method was incorporated. The difference between the old costs and the new lower

The Pareto Chart before improvements were attempted.

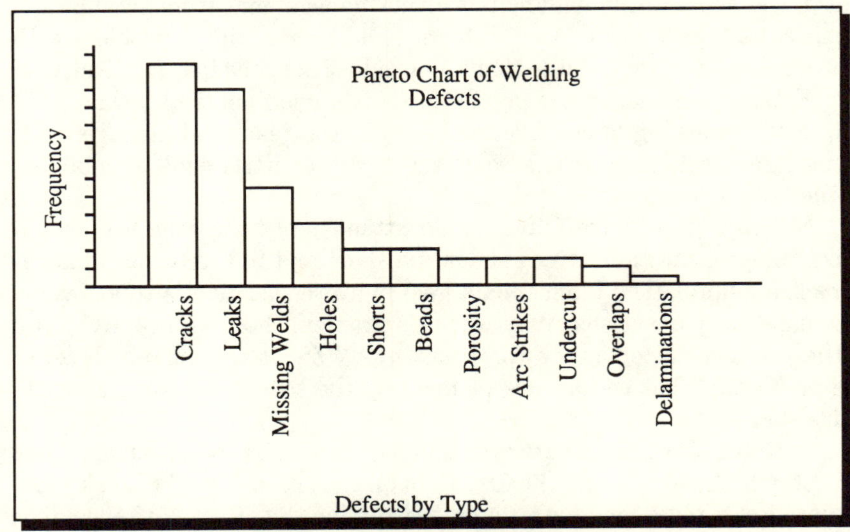

The Pareto chart after a single type of welding wire was used. Note how more than one type of defect was affected.

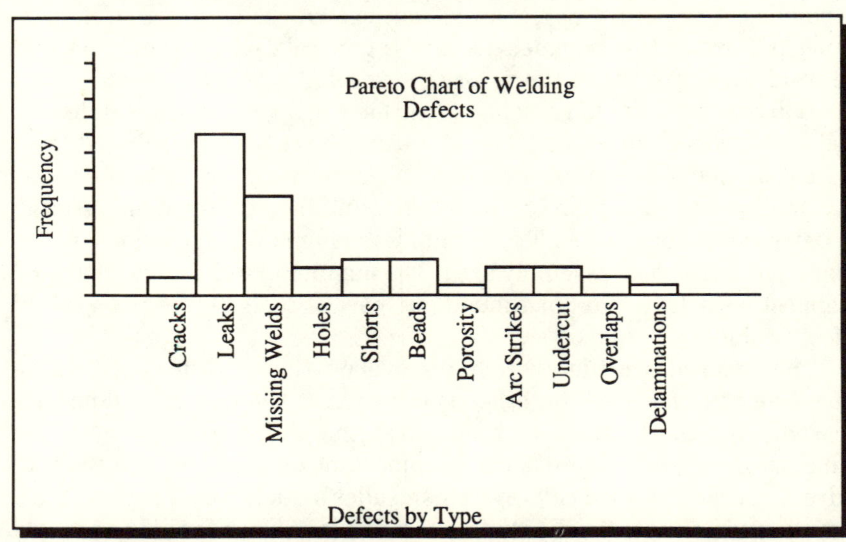

Figure 11.5 Using a Pareto chart to illustrate the amount of improvement.

Example of the calculation of the economic impact of an improvement.

Bob's work team attacked the problem of welding cracks by trying a single source of welding wire. Old wires that were used cost an average of $6.00 a pound. The new wire cost $7.00 a pound. However, the number of welding cracks found per day dropped from 10 to 3.

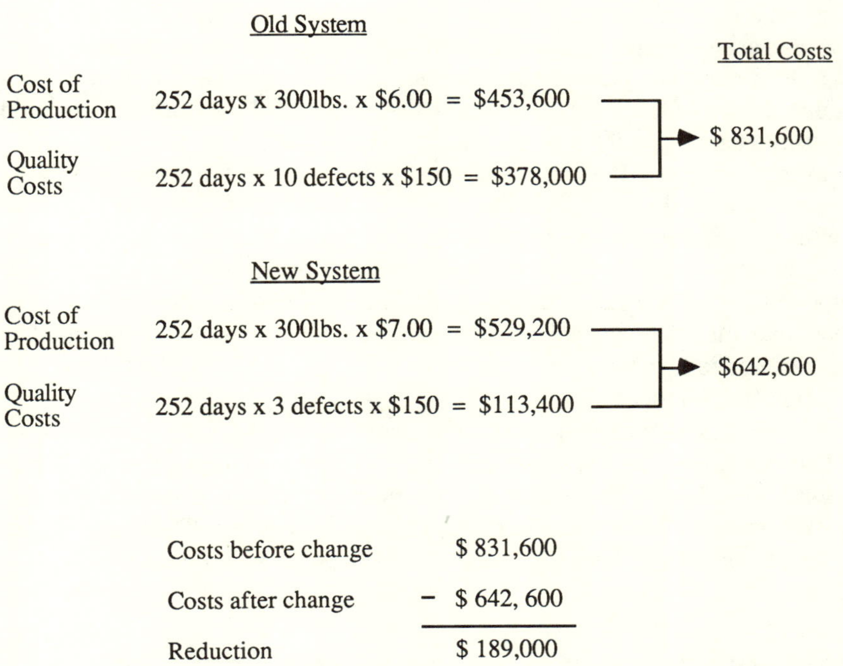

Figure 11.6 One way to calculate cost reductions realized from process improvements.

cost of fewer cracks occurring is the cost reduction realized by the corrective action. If greater than the additional expense of the wire, then continued use of the wire can be economically justified.

Step 7: Written Report for Management. The results of the actions taken, even the ones that fail, should be reported in writing to management. Successful actions can then be spread to other departments and processes. Attempts that failed can be compiled and published so that other work teams can avoid making similar mistakes. Again, this is the idea of exploiting opportunities and removing inhibitors.

The written report also helps management to evaluate a company-wide effect of each solution discovered. Management is an excellent vehicle for communicating results to the entire company. Other departments, such as sales, engineering, and purchasing, can learn from discoveries made by the teams. Bob's team's success may lead the purchasing people into reexamining the cost effectiveness of buying from multiple vendors. The written report enables management to better evaluate the performance of the people operating the processes. Conversely, the people operating the processes feel more involved with the company.

An important part of the written report should be the plan for implementation of the solution. Here the team details how the solution could be fully realized in its own as well as other related departments. Bob's team may suggest that buying from a single source for welding wire be tried in all welding areas. Also, the final report should include a list of inhibitors that were encountered by the team. Management's job is to review this list and try to remove as many as possible. Work teams may report that they could not find an engineer available who knew how to test machines for optimal process settings. The role of management is then either to hire such a person, or to train one of the existing engineers for this function.

Step 8. The important link. The problem-solving cycle ends with an important link—the never-ending link of continuous improvement. After a solution is found, there are other problems to attack. The team returns to step one and identifies the next problem to address. The cycle begins again. Every time the cycle repeats, the process is improved.

CONTROVERSY: HOW TO EVALUATE THE LEVEL OF SUCCESS.

The main controversy that a work team will face is the method used to evaluate each action. The above examples used a simple method of frequency and cost of each problem. Such a method could come under fire from the upper management of a company because managers tend to distrust financial information. The reason for this distrust is historic. Many companies suffer from "paper disease"—the need to justify themselves to stockholders by producing higher quarterly dividends. Managers at these companies have seen clever accountants manipulate numbers into attractive packages.

Bob's team demonstrates this kind of situation. It found that on paper the use of only one type of welding wire cut the cost of production by reducing the number of weld cracks. The key question is whether these reductions are real. Bob's team did not mention that the final assembly people had to take extra time to fill in a defect map which would cost extra money. Second, the team assumed that reduced weld cracks would mean less expense for rework. It is entirely possible that the rework people have so many defects to handle that eliminating this one failed to reduce their work load. Therefore, rework costs would remain constant.

Any problem-solving work team should keep this type of skepticism in mind when evaluating its success. Whether the fears of management are real or imagined, they should be addressed. The team should be able to respond to the questions raised by management. And most important of all, the effect of the team's effort should never be overlooked. Even if no cost savings are realized, the quality of the process and product is improving. The reward of improved quality is eventually realized by any company.

SUMMARY

Continuous improvement is a concept of dynamic involvement. Management, work teams, and individuals work with the idea of making each process a little better every day. To accomplish this objective, each person or group has to learn a little more about a process on a daily basis. The SPC system is the first step in the goal of continuous improvement. Control charts and process logs provide a wealth of information about the day-to-day activities of each process. Such information supplies a baseline that describes the typical day for the processes and each problem-solving effort can be evaluated against the changes in the baselines. The SPC system furnishes the information, but it is the people who make the changes resulting in the desired benefits. Management is forever changed. The manager who disseminates the day-to-day production decisions to the work teams, becomes both strategic leader and work team supporter. The key phrase in management in an SPC system is "decentralized authority."

CHAPTER 12
IMPACT OF SPC ON THE QUALITY DEPARTMENT

SPC redefines the roles and relationships of every department within a company, although the roles and responsibilities of the quality department are probably affected the most by the introduction of SPC. Tasks of inspection and monitoring the processes are assumed by the production personnel. The idea of quality control becomes obsolete in an SPC environment. To survive, the quality department becomes the central support group for the SPC system. The idea of quality assurance becomes a reality.

QUALITY CONTROL VERSUS QUALITY ASSURANCE

The terms quality control and quality assurance are not used here as departmental titles. Instead, they represent two different views of how the quality department should operate. Therefore, it is entirely possible that a department called quality assurance is really a quality control group. A brief description of each philosophy will help to clarify this difference.

A quality function is defined by its structure, purpose, and emphasis. A quality control group is structured to be part of the production function. Its main responsibility is to inspect goods as they are being produced, which sometimes includes both incoming and outgoing materials. A typical type of quality control operation is an inspector checking a large group of newly manufactured parts. A small sample is drawn and if too many defects are found, the entire group is rejected. The parts are sorted, scrapped, or reworked. The final decision about accepting or rejecting parts is usually left to a group other than quality control. The emphasis of the quality control system is to detect problems within the product.

The quality assurance group is structured as an independent department within the company. The function of the quality assurance group is to audit systems that assure quality of production. The information gathered by quality assurance is sent to all involved departments for their decision making. For example, the quality of parts in production are monitored by the operators of the process. Any parts found out of statistical control are isolated and quality assurance is called in to determine the actual quality of the parts. Production and quality assurance agree on disposal of the affected parts. The emphasis is on prevention of problems and reductions in manufacturing costs.

A complete SPC system requires a company to change from a quality control point of view to a quality assurance type of function, which is especially true since upper management is responsible for the control of quality under an SPC system. The idea of inspecting for quality is replaced by a management group who strives towards continuous improvement. Instead of rejecting bad parts or materials, resources are focused on finding ways to prevent problems, and changing roles and relationships of the quality department.

NEW ROLES FOR QUALITY PERSONNEL

The quality department replaces its former functions with three new roles it must perform to support the new SPC system:

1. Quality assurance as the facilitator and support group for SPC.
2. Quality personnel as auditors.
3. Quality assurance as a profit center for the company.

QUALITY ASSURANCE AS THE FACILITATOR AND SUPPORT GROUP FOR SPC

SPC, along with the philosophy of continuous improvement, requires that every person within the company should participate in the improvement of quality. To accomplish this objective, each person will have to be trained and supported. This function is best served by the quality assurance department. The quality assurance personnel traditionally are considered responsible for improvements and tend to have the most experience with using statistical methods. Therefore, well-trained quality personnel make excellent instructors and resource people for employees within an SPC system.

The quality and personnel departments must form a partnership in the training effort. Their first step is to list the goals of any training in terms of the skills they want participants to acquire from training. Potential participants must be indentified and screened by personnel for the math, reading, and writing skills they currently exhibit. Those with weaknesses will need special attention. The training content must always be relevant to the problem at hand. In the case of SPC it must include examples of the charts the participants will actually fill in. Finally, the participants must go quickly from training to actually applying the skills they have learned. Delay will only decrease the effectiveness of the training.

As a support group, quality assurance is responsible for providing technical assistance to any person or group using SPC. This could be something as simple as answering questions from line workers about the correct way to fill in a control chart or as complex as designing a new supplier quality assurance program for the purchasing department.

A key moment for any SPC system comes when the first operator using a control chart discovers a problem that is not easily correctable. An example can illustrate this point. Company Xeron has just begun to use control charts to monitor its production processes. Only a few days into the new program a screw machine operator shows up at the production manager's door with a valve lifter in one hand and a control chart in the other and shows the production manager that the chart indicates an average diameter is too large. The operator explains that the cutting tool is already set to the end of its stop and wants to know what to do. Obviously, the choice is either to stop and get a new cutting tool installed (losing two hours of production) or run the production anyway.

At this point, the statistical talents of the quality assurance department should be brought forth. The production manager should call a quality assurance person for technical advice so that the information on the control chart can be used to calculate the rate at which the diameter is increasing and how much scrap or rework can be expected if production waits to the end of the production run to repair the machine. In most cases, repair would be more cost efficient than a continued production run. The job may be done on time, but will probably involve costly sorting and reworking to assure the level of quality asked for by the customer. The quality assurance people in this example are not asked to approve the production run or to issue a "use-as-is" disposition. That layer of paper work and bureaucracy is stripped away. The production people make the production decisions, but also have to account for the quality of production. Thus, the quality assurance people become technical advisors to a production area held responsible for using control charts properly and forming problem-solving work teams.

QUALITY PERSONNEL AS AUDITORS

As noted in Chapter 6, frequent audits of the system ensure its effectiveness. In addition, all forms of inspection are replaced by periodic audits. The incoming inspection area is a good example of this change. A traditional role for quality control was to inspect incoming material using lot acceptance sampling plans. The idea was to screen out lots that contained too many defects. Under SPC, only zero defects are allowed, so inspection has to be replaced by a system that will encourage continuously improving quality of incoming materials. Purchasing plays a part in the system by encouraging vendors to pursue never ending improvements. The quality assurance department supports purchasing by auditing both supplier facilities and the quality of materials. Instead of accepting or rejecting a shipment, the quality assurance function is to audit the actual quality of the material. The information is then shared with the supplier and purchasing to help promote a dialog on improvement.

Defective materials will only create defective goods. This is the reason for the elimination of incoming inspection methods. Especially vulnerable have been the lot-by-lot sampling plans such as MIL-STD-105D. Under these plans, a company sets an acceptable quality level (AQL) for incoming materials. A typical AQL is 1.0%. Under such an established AQL, the company can expect to experience, on the average, 1.0% of the received material to be rejected. This system is quickly becoming obsolete because other systems such as SPC and JIT demand that all materials be defective free. The solution is to emphasize quality at the source. The earlier in the production cycle that quality can be assured, the more likely it can be maintained throughout production.

In a typical situation, the incoming inspection and testing personnel in a company are trained in quality assurance methods and teamed up with the purchasing group. Usually, the new team is called something like the purchasing quality assurance group. This group travels to the various sources of materials and services for the home company. Its mission is twofold. First, it must convince suppliers to adopt rigorous policies of continuous improvement to help improve quality from the source. Second, it must provide the technical support to the suppliers implementing new SPC systems. This method could also include advising suppliers with existing SPC systems on how to match their information to the needs of the home company.

The quality assurance department also audits to collect information for use by its own company. These audits are performed at three stages of the manufacturing process. First, the internal production operations are audited to assure that the SPC system is supporting continuous improvement (see Chapter 6). Next, the final products are occasionally audited as confirmation of predicted levels of quality. Any problems found with a final product are corrected by the production department. The role of the audit is for information only. Production, marketing and management are the main users of the information. The third stage audited is the actual use of the manufactured product. If the product is sent to an assembly plant, the auditors go there to measure how well the products work during assembly. If the product is sold to a final consumer, then the distributors, sales people and customers are interviewed. This procedure can be accomplished in several ways or combinations of ways. For instance, a market research group could sample consumer opinions about the product. The quality assurance people could examine customer complaints and warranty reports. Another possibility would be the face-to-face conversations with customers.

The information gathered is reported to top management by quality assurance and the same information is reported to the affected departments. The reports will include statistical summaries of the situations examined such as the quality assurance department's regular report on the Cpk indexes for all processes using SPC. In addition, all quality assurance reports must include cost information that is vital for helping management decide where to place valuable resources and people.

QUALITY COSTS VERSUS COST OF QUALITY

Reporting costs must be standardized within a company so management sees fair comparisons of problems or improvements. Adopting an accepted method of reporting quality related costs is a way to achieve this end. The two most popular methods are quality costs and cost of quality. The two should never be confused. *Quality costs* are all of the expenses a producer incurs while maintaining the quality of products and services. These costs are usually divided into four categories:

1. Prevention Costs. The expense involved in the prevention of problems that could affect the quality of the product or service. Example are cost of quality engineering, training, supporting problem-solving teams, et cetera.
2. Appraisal Costs. The expense of maintaining stated quality levels, including inspection, testing, and auditing.

3. Internal Failure Costs. The expense of correcting problems at the source of production. Scrap, rework, and repair are examples of these types of costs.
4. External Failure Costs. The expense of correcting problems after the product has been shipped or the service rendered. Complaints, warranty work, and lawsuits fall in this category.

The *Cost of Quality* (COQ) is seen as the cost of nonconformance. In other words—the cost incurred when things go wrong with production. The focus is on the added expense of maintaining quality while fighting problems. The COQ is usually reported as a single figure, such as the ratio of sales to the cost of nonconformance. Quality costs, on the other hand, concentrate on the wider field of all quality related operations, including those that are performed when no problems exist. For example, quality costs would track the salaries of the quality personnel, while COQ would probably ignore them as a fixed cost.

A controversy exists over how to use these four areas of quality costs to evaluate the cost effectiveness of quality improvements. On one side of the controversy are people like Phil Crosby advocating that all four categories should be reported as a single total. Then the effectiveness of improvements is measured as a reduction in the total cost of quality (COQ). On the other side of the controversy are people such as Armand Feigenbaum favoring separate detailed reports for each category. Improvements are evaluated according to their impact on particular sources of past expense and on the general effect on the quality costs. Instead of trying to reduce the total COQ, Feigenbaum recommends placing more money in prevention cost areas and directing efforts toward reducing costs in the other three areas.

For companies with a complete SPC system, the latter method is probably more useful. A successful SPC system will be evident by an increasing drop in internal and external failure costs. Later steps in implementing continuous improvement, such as elimination of incoming inspection, will also reduce appraisal costs. A good SPC system will have a cost accounting system that details the four quality cost categories by activity. For example, under appraisal costs, there would be accounting for the expense of inspection, setting up for inspection, time spent inspecting, test equipment depreciation, et cetera. Such information is useful in two ways. First, it allows upper management to examine the total impact of new problem-solving efforts. Second, it provides beneficial information for quality engineering. When bidding a new engineering job, the expense of assuring a high level of quality is known.

The COQ method of accounting for quality improvements is also of value in some settings. It is best for large companies that have many top executives. Since, the COQ is a single number that can be charted over time, the quality assurance department can communicate the improvement message more easily. Top managers tend to work with "the bottom line," while making corporate level decisions. The COQ is excellent for this purpose. Its other advantage is that it standardizes the measure of quality improvement. It prevents new managers from rearranging quality accounting reports to alter the true state of quality assurance costs.

Whichever method of cost accounting is used, it is best to pick one and stay with it. The changing role of the quality department is to make itself profitable. Since the department usually produces no product that can be sold for profit, it must help the other departments become more productive. In short, the quality department must become the facilitator of a more efficient company operation. It becomes the instigator and supporter of continuous improvement efforts.

QUALITY ASSURANCE AS A PROFIT CENTER FOR THE COMPANY

The above discussions about quality costs are important to the third function of a quality assurance department. The role of a well implemented quality assurance department is to increase the profits of a company by preventing defects, improving designs, and reducing production costs. In contrast, the quality control department is considered a cost of manufacturing. Inspection and rejection of materials are added expenses with very little pay back for the company as a whole.

To achieve continuous improvement, production forms problem-solving teams that use the information from the SPC system. Quality assurance's role is to provide technical support to these groups. Working as partners, the quality assurance people and the problem-solving groups will search for ways to improve the processes and thus, reduce operating costs. The Xeron example in Chapter 2 addressed this function. The team discovered that better sawing procedures almost eliminated the need for reworking shelves. Reduction in rework cuts the expense of production and directly increases the amount of profit a company produces making the quality assurance department a profit center for the company.

SPC and problem-solving teams will reduce quality costs a great deal in the short term from one to three years. Eventually the larger problems will be found and eliminated leaving smaller potential gains available for the SPC system. The long-term reductions in quality costs and increases in competitive position are realized by the continuous improvement of the design of products and processes. The Japanese have demonstrated this principle dramatically. A manufacturing plant in Japan is usually built on one-half of the land purchased for the building. After completion, the other half is the site of the plant to replace the first plant. From five to seven years after the first plant opens, it is closed and the second plant opens. The second plant is built using design improvements developed from the first plant and the original plant is gutted and newer designs are used in its reconstruction. The idea is to completely upgrade all operation every few years.

The quality people will have to work closely with both production and engineering areas in assisting with new designs. From the production area, the quality personnel will gather the control charts and process logs to summarize difficulties experienced during manufacturing. For example, they may find that resistance welding is a constant source of frustration for the work teams. The engineers can design out most resistance welding from specifications or they can investigate more effective welding methods.

NEW RELATIONSHIPS BETWEEN QUALITY ASSURANCE AND OTHER DEPARTMENTS

The most important relationship for quality assurance is with upper management. Quality is a company-wide effort under SPC and the quality assurance department is responsible as the main support and information gathering unit for the SPC system. At the same time, upper management needs a summary of this information for company-wide quality decisions. Quality assurance becomes a direct support group for management.

Conversely, when quality assurance is working with production, engineering or personnel departments, the relationship is a partnership. Gone are the days when the quality department must answer to the production manager or the chief engineer. Instead, the barriers among departments are removed so these groups can work as teams. The assurance of quality is part of any employee's job. The quality assurance people are available for technical support, guidance, advice, or extra manpower.

If the engineering group is planning to bid on a new part, the quality assurance people can assist by constructing a quality checklist for the bid detailing the quality assurance steps that would be taken during the production of the part, including the type and placement of control charts during production. In the production area, quality assurance would be continuously involved with the problem-solving teams. Quality assurance personnel are used as technical support. Usually, their role is to double-check capability studies and control charts to ensure their accuracy.

CUSTOMER RELATIONS

Another partnership that quality forms is with the sales and customer relations groups. Quality is again the pipeline of information about customer needs and the primary audience is management, production workers, and supervisors. The newest needs of the marketplace and the sales strategies of the company are important information for directing the efforts of the problem-solving teams. For example, an appliance manufacturer's quality assurance group may discuss with the production work teams that many dealers have been complaining about scratches in the finish of their appliances. Therefore, problem-solving efforts might focus on a minor defect, like scratches, that previously were not important during production.

Together, the sales, quality, and customer relations people can make joint presentations to upper management about the current state of meeting customer needs. The number and type of warranty claims made during the last 30 days could be discussed. Sales personnel could explain how the claims are handled; Customer relations would have background information on the nature of the claims; and quality could report on efforts being pursued by production and engineering in preventing the complaints. Of course, a really wise upper management group will be in constant contact with the customers anyway. Executives at Apple Computer must spend some time each month monitoring customer service calls as they come into the Apple Hotline. This procedure reflects the basis of quality and customer relations—the customer now drives the definition of "quality".

SUPPORT OF PRODUCTION METHODS

Zero inventories, Just-in-Time philosophies, or Computer Integrated Manufacturing (CIM) are just three kinds of new or developing production methods that have to be supported by the SPC system. Again, the quality department serves as the technical liaison between quality requirements and production methods. Zero inventories are a good example of this need. The idea of the zero inventory philosophy is that only the materials needed for immediate use will be kept on hand—usually at the site of production. To achieve this goal, production planners must be able to predict the capability of each stage of the process to a high degree of accuracy. The control charts and process logs are just some of the potential sources of this information.

FACTORS IMPORTANT TO PROFITABILITY

One difficulty about the quality assurance department demonstrating its cost effectiveness concerns the use of savings projections. A work team may discover a new method that will reduce the scrap rate on a part of the production line from 10% to 5%. The team then calculates how much this represents in dollars not spent for sorting or rework; usually, an annual savings figure is given. However, many times the anticipated savings are never realized because the production method changed or the correction was not fully implemented. Consequently, upper management will tend to lose confidence in cost reduction reports. For this reason quality cost reductions should be calculated conservatively.

A second part of this same controversy is the overall effect of cost reduction efforts. Many managers argue, and rightly so, that other important factors have to be considered. Productivity, effectiveness, and timeliness are also important measures of success. Therefore, a wise quality manager will account for these results in the quality cost report. Otherwise, problem-solving methods may be employed that might reduce scrap, but also reduce productivity.

SUMMARY

The introduction of SPC and other new manufacturing techniques have created a revolution in the function and definition of "quality". Gone are the days when quality was controlled by forming a separate department called quality control. It has been discovered that the least likely person to control quality is a quality control person. The emphasis today is to control quality by gaining statistical control of the processes. If the processes can be controlled, then quality can also be controlled. Manufacturing and management personnel must be responsible for the control of quality, and the quality control department must change or become obsolete. The required change is towards quality assurance. By supporting operations of other departments' functions in the SPC system, the quality department is a facilitator of SPC. By actively promoting problem-solving and cost saving methods, the quality assurance department eventually becomes a profit center for the company.

CHAPTER 13
COMPUTERS AND SPC

For the past few years the American Society for Quality Control (ASQC) has listed in its publication *Quality Progress* much of the software available for quality assurance functions. These programs assist in specific calculations that are done day after day. In that capacity they are very valuable, but, the very existence of an SPC system depends on accurate record keeping. Other types of software for personal computers can conquer some major record keeping problems. When used properly, a computer can cut costs, time, and paperwork from an SPC system. To accomplish this, the computer has to be selected and used carefully.

PURCHASING A COMPUTER

To avoid purchasing equipment that will soon be obsolete and to ensure the maximum use of the software, attention should focus on careful planning. A computer will magnify the effect of any system. In other words, a well-designed, smoothly run paper system will usually gain in efficiency by adding a computer. For poorly designed paper systems, the computer will increase the number of errors and frustrations. This principle is all important in the selection of both computer hardware and software.

Step 1: Describe the mission.

More simply put, what is the purpose of obtaining a computer to assist in SPC operations? Many companies create this mission statement by naming the people they expect to use the computer and the type of activities in which they will be engaged. Such a plan would say something like, ". . . the production supervisor will use the database program to enter control chart notes at the end of each day . . ." Another approach is to describe the functions available on the screen for any user. For example, ". . . the main menu will include options for data capture, chart creation, capability study, and use of the process log."

The intention of such a mission statement is to establish a clear picture of how the proposed computer equipment will be used, which will aid in the selection of software and hardware that will meet expectations.

Step 2: Select the Software.

Before looking for the computer, maximum time should be spent selecting the computer software that will perform the stated mission. This step is critical because it is the software that will actually do the work. Software is the set of program instructions that tells the computer what tasks to perform. The computer is a slave to the program instructions.

When selecting software, it is advisable to test all packages with actual examples of the information that will be used at the host company. The computer industry is still young and prone to problems, such as "vaporware." Vaporware is software that is advertised or promised but in reality will never see the marketplace. Therefore, the software purchaser should accept nothing that is not physically demonstrated to exist. A program that is difficult to use or has errors must be rejected, even if the sales people promise it will soon be fixed.

Once the appropriate software is selected, the job is to find a machine capable of operating all of the programs.

Step 3: Selecting the Hardware.

The key to choosing a computer is to find one that will run the selected software. One method is to list all of the software packages needed and the computers with which they are compatible. If one machine can run all of the packages, then the choice is obvious. On the other hand, if not one machine can run all of the software, then a more exacting task is needed. The company must test alternative packages that will fill the gaps in software packages. However, many companies have found that common personal computers such as the IBM-PC, Hewlett-Packard or Apple can run their selection of software.

The needs of a quality assurance department can illustrate how to select hardware. If the quality assurance manager has decided that one personal computer should be stationed in the department to handle all SPC related information, the computer will need a powerful database software package to store and retrieve thousands of records containing control chart information. The machine will also do statistical calculations, weekly reports, capability studies, and some correspondence with vendors. The personal computer would include the maximum amount of RAM memory chips and a math coprocessor chip to speed calculations. A fixed disk would be added for storing the SPC information. A letter quality printer would be used for correspondence and reports while a dot-matrix printer would produce quick capability studies. Finally, a tape backup device would be necessary for a backup copy of stored data files so that all the records are not lost if the computer should fail.

Step 4: Training the Users.

Training is one expense most purchasers overlook—how will the people who are to use the computer learn how to operate it? Two levels to the training are needed to make people proficient with computers. Level one is the basic operation of the computer. Topics include how to turn it on, run the printer, how to use the disk operating system (DOS) and how to load software programs. Level two is the actual operation of those programs, and usually requires a series of seminars or courses. A specific package is emphasized in each course. One course could be on word processing where the participants learn how to operate the word processing package and see how it is far more efficient than typing. Community colleges, computer stores, and private training groups usually offer experienced instruction in these subjects. Such training should be for the specific software used on-site and include the use of spare computers for the participant to practice on between sessions.

Proper computer training can greatly enhance the value of a small computer system. One of the main complaints from companies using personal computers is that they sit idle most of the time or that little productivity is seen. Both of these conditions can be easily corrected by training a large group of people in computer use. Like anything that is new and different, people are hesitant to adopt it. This is the reason "user groups" are so popular. They are a collection of computer users who meet regularly to discuss new procedures and techniques. Such an environment makes a novice feel more at ease in asking questions.

A common mistake made by many first-time computer purchasers is thinking that the cost of the computer is the biggest expense involved. In reality the cost of the computer is one of the lowest expenses. More money is needed for the proper software and training. Without good software packages and well-trained employees, the computer can quickly become an expensive paperweight.

COMPUTER SOFTWARE THAT SUPPORTS AN SPC SYSTEM

As mentioned before, numerous software packages can assist in the operation of an SPC system. Dozens of companies produce software that can create control charts and capability studies based on raw data the operator enters. Such programs are a great benefit to any overworked

SPC coordinator who must post several new charts every week, although these programs are valuable only to the people directly involved in the creation of charts.

Related software programs are the statistical programs which provide excellent support for companies that need to do detailed statistical summaries of processes and finished products for internal and external use. These programs can perform statistical calculations hundreds of times faster than a person can with a calculator. In addition, many companies select such a program along with a computer that can be connected to their coordinate measuring machine to help speed first-part production tests, gage verification, and other related activities.

A family of programs for general quality assurance functions exists and most of these contain SPC components. They also have features such as lot-by-lot sampling plans, statistical summaries, lot traceability, life testing, gage studies, and remote data capture. Some of these programs can be extensive, and some allow several computers to be linked together from sites around the factory to share information. Thus, a part can be tracked at each step of the process and a quality report made available from any of the computers.

When selecting software, a couple of points should be kept in mind. As always, a working demonstration of the software should be demanded. Second, the company producing the software should be well-established. Many times a good product is sold by a company that will go out of business within a year leaving the company with no support or improved versions of the software. Usually, well-established companies have existed for several years and have a long list of customers. Third, the software should be checked for expandability. Are there other modules that can be purchased to increase the number of applications the software can handle?

Other types of software available can greatly increase the efficiency of an SPC system by helping to store, sort and present information. Most of these programs are intended for conventional office applications, however, they can also support the information flow created by an SPC system.

Word Processing. Word processing programs are much more than electronic typing. New packages include dozens of aids to increase writing efficiency. Spelling check programs will seek out and change misspellings, mail merge functions can create several individualized reports from one document in seconds, and some programs allow the production of multiple columns of print for more professional communications. All of the packages allow the creation of standardized report forms that can be quickly modified.

For an SPC system the first suggested application for a word processing program is creation of the reaction plans for the control charts. As stated in earlier chapters, the reaction plan is a dynamic document. The word processor can be used to draft and save the first reaction plan for a process. As experience requires modifications of the plan, it can be called up on the screen, quickly changed, and printed out for use on the floor. The same procedure can be used for lists of charts being implemented, management's strategic plan, measuring standard, and quality control inventories.

Database Programs. A database program is designed to store information in an orderly fashion for easy retrieval and summary of the information. In fact, it is this retrieval and summary function that should be most closely examined. An example will illustrate why. A typical database will allow storage of information in "fields", such as product name, number, average size, standard deviation, et cetera. Once entered, the thousands of records can be searched and sorted into a variety of reports. A quality audit team may enter all warranty information, then using the search capabilities of the program, it can sort by region, complaint, part involved, and date. This search could reveal that during the summer months, rust is a large problem in the southeastern region. What the program detected and reported in a few minutes the team can investigate. The most probable cause is high humidity in the Southeast during the summer. Since, the computer can conduct searches in seconds, several dozen "what-if" scenarios can be tried in one day. Before this capability was available, months of record searching would be required for a single test.

Spreadsheets. Any one who has studied accounting is familiar with the spreadsheet. It's a grid of rows and columns for setting up financial models. Some industrious accounting teachers developed the first electronic spreadsheets. Their intention was to have the row and column totals automatically recalculate every time a change was made on the spreadsheet. As it turns out, the electronic spreadsheet was widely successful for more than accountants, especially for SPC.

Most statistics are calculated by hand using rows and columns of data. It is fairly easy to set up statistical spreadsheets with these programs. In fact, the newer spreadsheets have built in functions for averages, standard deviations, and linear regressions. Therefore, a trained operator of a spreadsheet program could create "templates" for calculating capability studies and control chart information. All that the operator would enter is the raw data.

The spreadsheet also has a more obvious financial application. The cost of quality model used by a company can be entered, copies of the

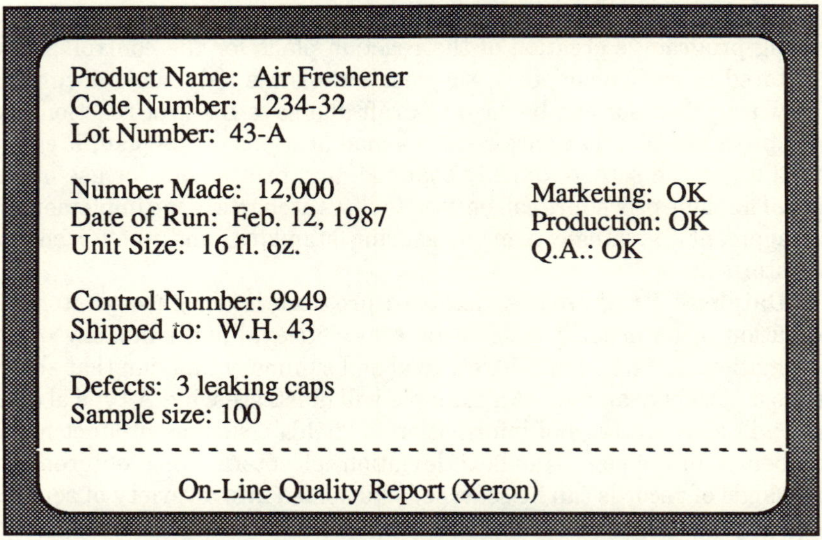

Product Name: Air Freshener
Code Number: 1234-32
Lot Number: 43-A

Number Made: 12,000   Marketing: OK
Date of Run: Feb. 12, 1987   Production: OK
Unit Size: 16 fl. oz.   Q.A.: OK

Control Number: 9949
Shipped to: W.H. 43

Defects: 3 leaking caps
Sample size: 100

On-Line Quality Report (Xeron)

Database as it appears on the computer screen.

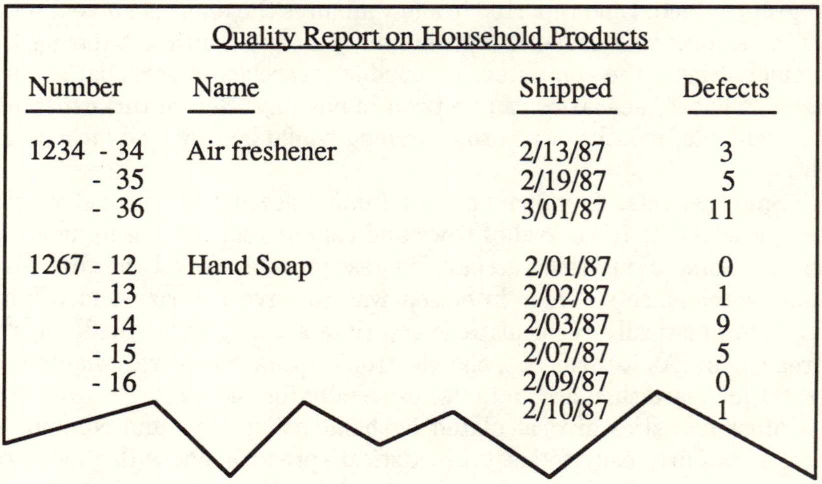

Quality Report on Household Products			
Number	Name	Shipped	Defects
1234 - 34	Air freshener	2/13/87	3
- 35		2/19/87	5
- 36		3/01/87	11
1267 - 12	Hand Soap	2/01/87	0
- 13		2/02/87	1
- 14		2/03/87	9
- 15		2/07/87	5
- 16		2/09/87	0
		2/10/87	1

Just one type of report that can be generated from a single database.

Figure 13.1 Example of functions an electronic database can perform.

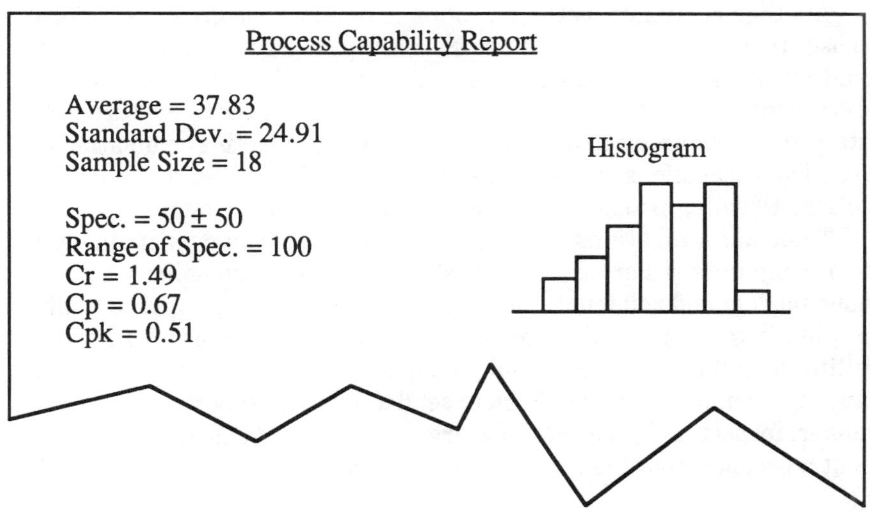

Electronic Spreadsheet on a computer screen.

Process Capability Report

Average = 37.83
Standard Dev. = 24.91
Sample Size = 18

Spec. = 50 ± 50
Range of Spec. = 100
Cr = 1.49
Cp = 0.67
Cpk = 0.51

Example of one of the reports that can be generated by
an electronic spreadsheet.

Figure 13.2 The use of an electronic spreadsheet to make SPC related
calculations and reports.

model can be made for each process, and information on specific costs for the process can be entered and printed. Every month or quarter, improvements are entered and printed to compare to the previous report. In fact, some spreadsheet programs create these differences on each report automatically.

Other possible applications include financial models, simulation studies, and capacity planning formulas. There are actually hundreds of uses for the spreadsheet in an SPC system, but one more will be mentioned for illustration. The capability indexes from each product and process can be listed on a spreadsheet and the sheet can be sorted electronically. The result is a list of processes from the best to the worst capability ratings, creating the company's "hit list" of processes that need attention. As could be imagined, the type and number of defects plantwide could be put through a similar process.

Custom Programs. Several programming languages are available for computers that allow a skilled programmer to create custom programs for the company. This procedure is time-consuming and expensive, so it is recommended only for highly specialized applications. An example would be the creation of a special type of chart for real-time use on the floor.

Another possibility is the hiring of computer consultants who have licenses to modify over-the-counter software packages. Such a consultant could modify a spreadsheet package so that it could obtain information directly from measurement gages allowing the operator to capture SPC data directly to an electronic spreadsheet for conducting simulation or capability calculations. However, such modifications must be used with caution. Usually, modifications will void the software's warranty.

Telecommunications. A computer is capable of communicating with other computers either directly by cable or over telephone lines. A telephone modem and communications program allows one computer to dial up other computers or databases hooked to other phones. An auditor visiting a supplier in Texas could sample parts and dial up the home company's computer in New York to send a complete report. In the same manner, inspectors in the back of a large factory could call up information about a product from the computer located on the other end of the building.

Direct connection is the method of attaching a computer to the company's mainframe computer or computer network. Using this technique, it is possible to download information from the main computer for use on the personal computer. Once the information is obtained, then the communication can be turned off saving valuable mainframe time and use. By the same measure, the mainframe computer can be used as a central information base for the whole company. Electronic mail and database sharing allows widely separated groups to communicate with each other at the same time.

The telephone dialing capability is another important function. It allows any personal computer user to access any of the hundreds of electronic databases located around the country. Dialog is an "encyclopedic" database run by the Lockheed Corporation. On this database are the indexes and abstracts of thousands of publications. A company may want to know what the Japanese have done in the way of applying for patents for three-way value designs. By calling up one of these storehouses of information, a user can search for patent information from several countries. However, these services are expensive. Some searches cost as much as $120 an hour and up, so only a skilled database searcher experienced with the service should be employed.

One other use for telecommunications involves electronic conferences where a central computer connects several companies through phone modems. Each company can dial up the computer and enter into an on-line conference with other users. Several of these systems currently exist in the United States to share information about SPC technology and implementation.

Remote Data Capture. One related piece of equipment is a combination of computer and software in a small portable form and is grouped under the generic title of remote data capture devices. The DataMyte and Gage Talker are two such products that are usually smaller than a textbook. One or more measurement gages can be connected to them. The device enables a person to enter readings directly from a measurement gage to the capture device. After making a series of measurements, the data capture device summarizes the results. Most produce a simple statistical report but many can also calculate data needed to create control charts. One important feature is the ability to feed the captured information into a personal computer.

Incoming inspection is an application where the remote data capture device is particularly well-suited. For example, a quality auditor can attach a caliper to the device and freely walk among the piles of incoming materials to collect data. At each shipment, the auditor can randomly sample some of the parts and arrange them on a flat surface. Using a measurement gage attached to the device, readings can be made and recorded quickly. No paper work is necessary. After inspecting several lots of parts, the auditor returns to the office to enter the captured data into the personal computer. Statistical summaries and capability studies are quickly produced. The entire process is done in minutes instead of hours.

Computer Aided Design (CAD). Currently, General Motors is planning to convert all blueprints into CAD which is one example of the move towards a world of electronic drawings. SPC and quality people should consider the use of a CAD package on their computers. One possibility would be the creation of critical dimension illustrations for the control charts by editing the electronic drawing. Another possible use is the review of new drawings for quality features. Notes and comments could be attached to the drawings' database area.

Several companies have discovered an interesting use for CAD packages with SPC systems; they design their own control charting paper using their CAD system. This procedure allows the company to custom design each control chart to the application within minutes and control limits or special note areas can also be added to the chart.

FUNCTIONS OF A COMPUTER

The computer must always be the slave of the SPC system. For example, one company placed mainframe computer terminals at each stage of the production line. Each operator could call up the right control chart to fill in, could enter process log information and scan specifications, while also having access to other production information such as the bill of material.

Other companies have tied personal computers to the measurement devices used in the SPC system. In one instance a company had to weigh five parts every half an hour for an X-bar/R control chart. The scale used was connected to the computer. All the operator had to do was to place each part on the scale, step on a foot switch, and the computer would fill in the chart.

It is strongly suggested that any company wanting an automated SPC system first use the paper system. By filling in charts by hand, the operators, supervisors, and managers learn how to interpret and use SPC information. This "hands on" approach reinforces the learning process. Only when the employees are comfortable with using SPC information, should the system be automated because very little is gained by having a computer produce reports that few people can read, let alone use.

An advanced stage of combining computers with an SPC system is to have the computers form a feedback loop with the process. At an automotive company that ground parts for engines, the computer was first employed for capability studies on the amount of variation experienced for a critical dimension. This information was used to predict how well the part would assemble to its mating component. Eventually, the process was changed so that the computer sorted the parts into exact sizes. Groups of specific sizes would be mated to the components that they fit exactly which eliminated the effect of part variations.

COMPUTERS MAY REDUCE HUMAN CREATIVITY

There are several controversies associated with computers. Some claim that the computer sorts out people who cannot cope with technology well and puts them at even more of a disadvantage for finding jobs. However, one issue is specifically related to the principles of SPC and that is, the computer tends to stifle creativity and innovation. The computer is a calculating machine; it sees no shades of gray in a situation. When SPC information is stored on a computer, it can become fixed and unchallenged. This condition will tend to make the SPC system resemble regulation instead of an opportunity for improvement.

Information must be seen as just that—information. Data is only evidence; it is not proof. The final answer lies in the decisions made by people. Consequently, the ability of the computer to summarize quantities of information must be used by the problem-solving teams to find creative solutions to identified problems. The key question is "what if?"

SUMMARY

The increased flow of information spawned by an SPC system all but requires the use of a computer. If for no other reason, a computer will speed the calculation of capability studies and control limits, but there are also several other good reasons for using a computer with an SPC system. First, the use of database, spreadsheet, word processing, and telecommunicating software expands the utility of the SPC system, especially the quality department's capacity to provide technical support. In addition, the SPC software package can be a great aid in making charts, conducting studies, and testing new methods. Other tasks for the computer to perform include gathering data from a remote location or operating controllers within the process. In short, the list of capabilities for a computer in an SPC system is still being written.

Of course, making the computer a valuable addition to an SPC system requires careful planning. The selection of software and hardware has to be preceeded by a statement of the machine's mission and intended use. Operators have to be trained in its use and potential. Otherwise, lack of understanding about the computer will create an atmosphere of fear and eventually, the disuse of the machine.

CHAPTER 14
SUPPLIERS, CUSTOMERS, AND COMPETITORS

A company's relationship between suppliers and customers and competitors is currently based on legal contracts. The legal system in the United States is designed as adversarial, so, it is not surprising to find that supplier, customer, and competitor relationships have also tended to be adversarial. Legal contracts drawn up by these parties have been seen as protection from one another. All conditions of any business agreement had to be spelled out in great detail, otherwise one party might take advantage of the other.

In the past, it was not unusual to find a buyer who had an elaborate system of incoming inspection for purchased materials. The purpose of the incoming inspection unit was to detect any materials that did not meet the company's specifications. Such inspections are expensive to conduct and even more expensive when they discover unsatisfactory goods. If incoming inspection frequently finds faulty material, the company must go through the expense of notifying the supplier and disposing of the rejected material. Usually the material is returned to the supplier at additional cost.

A company implementing a complete SPC system can escape the above scenarios. SPC can be spread to the suppliers so they are a source of quality, instead of a cause of defective products. This condition is accomplished by establishing clear lines of communication with suppliers which begins with clearly understandable quality standards based on the hard numbers required by SPC.

Today a minor revolution is occurring in the United States about the ways different businesses treat each other and is based on the concept that cooperation is the best competition. Companies are redesigning supplier contracts to encourage sharing knowledge and opportunities. Businesses that were formally competitors are entering into joint ventures and are sharing both effort and profits.

CHANGING ROLES OF CUSTOMER AND SUPPLIER

Most companies are suppliers in one way or another and provide goods or services to other companies. All companies are customers and receivers of goods and services, meaning they are usually both customer and supplier at the same time. Some adapted to this situation by creating a supplier quality assurance program to complement their purchasing agreements. When goods or services are purchased, the contract contains information about the expected level of quality. The supplier was expected to meet these specifications. The supplier quality assurance department did not have enough people to help the hundreds of suppliers. In addition, incoming inspectors would carefully examine each shipment of products to confirm its quality before the company completed the purchase.

From the supplier's perspective the rules seemed inconsistent. Quite often, a company's policy about purchasing had no resemblance to the policies about its own production. Manufacturing was geared to creating products that would just meet the customer's requirements. In fact, for a long time in the United States the definition of "quality" was meeting customer requirements. There was no thought of trying to exceed requirements or of taking the time to make a process more cost efficient while production was running. If a lot of parts was rejected by the customer, it wasn't unusual for the supplier to sort a few bad parts out and send the same shipment back.

In short, the relationship between customers and suppliers was sometimes a game of deception. The customer would try to enforce strict conditions it would not impose in its own plants while the supplier tried to spend as little time and money as it took to meet minimum requirements. This was the general situation until recently but an economic recession changed thinking about customer/supplier relations.

The recession of the late 1970s made expenses more painful. The high cost of inspection versus income for companies was noticed by upper managers. In addition, three additional factors made managers aware of the high cost of established quality techniques. These three factors were: product liability, warranties, and recalls. Several landmark cases in the 1970s increased a producer's responsibilities for product liability. The case of Micallef vs. Miehle Co., New York (1976) found that a company was liable for an injury caused by a product if the company had the reasonable ability to design out the danger, even if the person injured knew about the danger. Many noted liability suits against companies followed. In addition, the companies were still responsible for honoring warranties on their products or recalling defective products. In short, poor quality was very expensive for any company.

At the same time, individual consumers wanted more value per unit purchased. Money was tight and customers needed low cost, high quality products that were reliable and durable. To manufacture such products and to avoid liability, warranty, and recall expenses, production companies also began to demand higher quality from their suppliers which led to a revolution in supplier relations.

As many case studies illustrate, the relationship between a purchasing company and a supply company is changing. Several types of supplier programs have emerged in the past few years, but all seem to adopt common principles.

New Awareness of a Demand for Quality

The new supplier relationships begin with new requirements and contracts that emphasize the pursuit of continuous improvement. Buyers had recognized the new market demand for quality and sent a single clear message of this need to the supply companies. As will be seen below, this began with the requirement for a tangible system of Quality Assurance, in most cases, SPC. After the principles of SPC were introduced to the supplier, more intangible concepts, such as continuous improvement, were easily introduced.

Emphasis on Quality from the Source

A buyer must make the supplier aware of the benefits that both companies can achieve by cooperating on quality. Of course, the obvious result of the new supplier requirements is that the supplier will be responsible for more quality assurance functions at its own facilities. This has to be counterbalanced by the buyer pointing out the cost savings for the supplier from continuous improvement. An SPC system at the supplier's should catch problems before products are shipped to the buyer and rejected. Some buyers sweeten the relationship by offering long-term contracts and shared technical resources for cooperative suppliers.

Awareness of Liability and Warranty Responsibilities

Another highlight of new supplier agreements is a renewed interest in thorough testing of new products before production begins. Capability tests and engineering studies are required before production is approved for the supplier. Producers recognize that liability and warranty costs can be disastrous when products are not carefully designed and tested.

Effectiveness of Cooperation Between Buyers and Suppliers.

When a buyer and a supplier have a complete SPC system, a high level of cooperation is possible between the two companies. To achieve this degree of cooperation the relationship has to be personal and direct. One person at each company must be identified to be responsible for all communication between the two companies.

Total Quality Control (TQC).

The essence of TQC is that every person at a company is in some way responsible for the creation of quality in a product or service which develops the concept of quality as a method of management. It has been argued in previous chapters that SPC and continuous improvement form a complete management system for a company. The following case studies reveal that General Motors and Ford Motor Company also feel that a change in management techniques is a supplier requirement.

Elimination of Incoming Inspection.

Although not spelled out in many new supplier contracts, many companies would like to phase out their incoming inspection areas. Long a source of expense, the incoming inspection process was doomed after American managers discovered that many Japanese companies have no incoming inspection. Instead, the Japanese spend many hours at their supply companies closely monitoring measurable traits of quality in the processes that helps to locate and eliminate problems before products reach the purchasing company. In addition, knowing that a supplied product is of high quality means that materials can be shipped straight from the supplier to the assembly lines. The smooth flow from supplier to buyer facilitates Just-in-Time systems of delivery. Needless to say, things like acceptable quality levels (AQL's) and lot sampling plans become obsolete in such an environment.

CASE STUDY: GENERAL MOTORS' VISION OF THE FUTURE

The major automotive manufacturers were hard hit by the recession of the late 1970s. They were struck with three major changes in the industry at the same time. First, the oil crisis created a demand for fuel efficient cars. Second, the Japanese made major inroads into American markets. Third, the recession reduced the ability to adapt to the above two conditions. One step that all domestic automakers took was to study the Japanese automotive industry. A particular point, noted by Ford and General Motors, was that, supplier relations should reflect the company's operating style. In other words, the construction of a car should be under one operating philosophy including everything from mining raw materials to sale of the car. Suppliers and the assembly plants must operate under the same system. Suppliers must be involved partners with the automotive companies.

Nissan, Toyota, and Mazda had shown that a high level of cooperation between suppliers and purchasing companies was extremely cost effective. Techniques such as Just-in-Time deliveries, Kanban and statistical process control were making it possible for the Japanese to build a car for about $1,700 less per copy then it would cost to build the same car in the United States. Many of these techniques and methods were studied by the U.S. automakers. U.S. automakers returned home to scrap their old vendor relationship policies and to adopt an entirely new view of supplier relationships. General Motors has even published a statement of its vision of future relationships with suppliers.

1. Continuous Improvement. General Motors' operating philosophy specifically states that continuous improvement will be the policy for all business relationships. For the suppliers, this means that GM expects them to also have a policy of continuous improvement.
2. System Suppliers. For greater efficiency in final assembly of a car GM is creating system suppliers where feasible. Instead of producing just parts, system suppliers create a single system for the car. An example is the control panel. Suppliers would be encouraged to produce the entire panel instead of only furnishing a few parts for the panel. The reason for this method is to place the concern about the interrelation between parts into the hands of a single company, namely the supplier.
3. Early Involvement. Instead of awarding contracts once a new part is designed, GM involves suppliers much earlier in the process. In the use of top rated suppliers, GM allows them to design and develop new parts. This procedure is intended to ensure the manufacturability of the part.
4. Certification. Suppliers for GM are certified, qualified or both. The quality and production systems at the supplier are closely inspected. Much of purchasing's quality assurance role is centered on encouraging suppliers to adopt operating systems similar to those used at GM. A Ford Motor Company executive said it best when he said that "the president of a supply company is

becoming the first line supervisor for Ford." In other words, supply companies are becoming part of the purchasing companies' assembly line.

5. Team Work. GM is actively promoting the idea of participation between suppliers and GM and is sharing expertise and information with its suppliers. Suppliers are allowed to voice opinions about designs and specifications. In addition, GM is sending technical experts to the suppliers to help implement new technologies. This is a radical departure from past practices. Cooperation has replaced adversity. Today, it is not unusual for a first rate GM supplier (or Ford for that matter) to receive a part design with a fair price to be agreed upon later.

Ford has also developed supplier relationships similar to those above and both GM and Ford have adopted one technique that has had a dramatic impact on automotive suppliers—the single source policy. The plan is to have only one supplier for each basic component or part purchased. Multiple sourcing of parts is being eliminated. This procedure is expected to reduce the number of automotive suppliers by up to two-thirds. The auto makers' intentions are to eliminate duplication of effort and disagreements on interpreting specifications.

CASE STUDY: FORD MOTOR COMPANY'S QUALITY SYSTEM STANDARD

Ford Motor Company modified its supplier quality requirements in the late 1970s. These requirements were published in a Ford manual called *Q-101*. Q-101 lists the quality standards a supply company must meet or exceed to retain Ford contracts. The new Q-101 represents a system of operation that more closely matches the manufacturing improvements planned within Ford. It also changes the system for ranking suppliers. To achieve Ford's highest supplier rating (Q1), a company must have a documented and demonstrated system of quality assurance and a flawless record of part quality. While GM allows some flexibility in the quality systems used by suppliers, Ford is rigid. Only one system is allowed and all Ford suppliers must have it.

Below is a summary of some of the parts of the Q-101 system. It is important to note that the system is designed to help supplier procedure match the operations inside Ford plants. The intention of Q-101 and its related purchasing policies is to make the supplier an integrated part of the Ford Motor Company. The importance of the Q-101 for study is that many nonautomotive companies in the United States have studied or copied Q-101 for their own supplier quality assurance programs.

1. Strategic Planning. Each supplier is encouraged to engage in long-range strategic plans for its company. The strategic plan should address how the company intends to remain competitive while expanding the number of services and parts it offers.
2. Continuous Improvement. Just like GM, Ford suppliers are expected to adopt a philosophy of continuous improvement.
3. Control Plans. The production of each part for Ford must be preceded by the submission of a quality control plan that is a summary of quality assurance procedures affecting the particular part.
4. Process Capability Studies. Suppliers to Ford are required to study the capability of processes to produce the proper fit, function, durability, and appearance of manufactured parts. These capability studies are divided into two general types. The first is the process potential study that makes a preliminary assessment of the process' potential in meeting specifications. Usually this is the 30 piece capability study required for contract bidding. The second step is the process capability study. In this step, statistical control of the process is achieved, then capability evaluated. (Chapters 2 and 3 are based on this model.)
5. Statistical Process Control. SPC is designated as the key system for providing information necessary for the continuous improvement of production and quality. All processes at a supplier's plant must be under an SPC system; any exceptions must be negotiated with Ford. The SPC system is critical for Ford. When inspecting a supplier or tracking down root causes of a production problem, the SPC charts will be the first information requested.
6. Ongoing Verification and Inspection. SPC is not the only method for assuring the quality of parts. No SPC system is complete by itself, so other verification methods and extensive testing are required. Many automotive parts are subjected to salt spray tests by exposing them to a salt spray for days at a time. Obviously, the information from this test cannot be used on control charts. Instead, such testing and inspection results must be reported in a timely manner to Ford.

     Ford also requires periodic layout inspections of its parts

and related gages. Periodic audits are also expected which cover the production process and the SPC system. In addition, any product returned from Ford must undergo extensive analysis for the causes of described problems. Ford then expects a report on the corrective actions taken by the company.

7. Documentation. All procedures, changes in the production process, and control of materials for production must be documented. An example is the requirement for the control of drawings; all blueprints issued by Ford to the supplier must be available on demand. In addition, if a current print refers to other Ford documents, these documents must also be readily available. Obsolete blueprint must be accounted for in writing and destroyed. If the supplier requests a change in a print, it must be agreed to in writing from the Ford purchasing department.

8. Ford SQA Surveys. Ford, like many companies, has a Supplier Quality Assurance (SQA) group. These people travel to the suppliers to inspect their facilities and production systems. The important distinction about the Ford SQA groups is that they are particularly interested in the system for assuring the high quality of the supplied part. Before Purchasing will award a contract, the SQA group will inspect the supplier. In addition to these duties, the SQA group also requires suppliers to make trial runs of new parts. The usual procedure is to have the supplier produce 300 parts on its production equipment. Note that special runs or hand finished pieces are not allowed. Ford's SQA group wants to evaluate the effectiveness of the suppliers' actual production equipment. Usually, this is done by statistically studying 50 randomly selected parts from the trial run of 300 parts.

The above procedures are designed to help the supplier save money and operate more efficiently. To many supply companies, these benefits are not obvious upon reading Q-101 because the benefits are usually achieved after a long period of implementation and development. However, the benefits have been quite real for many Ford suppliers. Those suppliers that win the Q1 award typically have found immediate economic benefits. Primarily, Ford gives such companies first consideration for new part contracts. The first sample evaluations, the submission of control plans, and the SQA surveys are discontinued. Ford is essentially indicating its trust in the supply company. Internally, the philosophy of continuous improvement has allowed cuts in production costs. Q1 suppliers have reported additional benefits from outside Ford—the publicity of achieving Q1 has resulted in winning unsolicited contracts from new companies.

IMPLEMENTING A NEW SUPPLIER RELATIONSHIP

In the past, a typical supplier only heard from a buyer when a new contract was offered or when a shipment was rejected. New supplier relationships demand a constant stream of communication between suppliers and buyers. The buyer has to require more quality assurance from the supplier. However, the buyer can make additional efforts more attractive by offering greater benefits to the supply company. Long-term contracts, shared resources, and no incoming inspection are just some of the advantages a buyer can offer a supply company. In addition, a buyer can offer free supplier training courses that would bring suppliers to the buyer's company and explain the benefits of the new relationship and new technologies such as SPC. Such training is enhanced when the buyer's own production lines are using SPC and the participants are able to tour the facilities. The buyer demonstrates the advantages of SPC first hand.

Of course, creating contracts with elements like those in Q-101 is part of the new relationship. Such contracts will be the supplier's main interest. If the new requirements are clearly communicated along with the benefits, then cooperation should be accomplished easily. The message is simple to express, "You can't inspect quality into a product." The buyer must trust and cooperate with the suppliers to achieve higher levels of quality. More quality means more customers and orders for mutual gain. Supplier relations must always be a "win-win" situation.

COMPETITORS

Along with dramatic changes in the relationships between buyers and suppliers, there is also a massive shift in the relationship between companies that compete. These changes have been marked by take-overs, joint ventures, and underselling. The end result of these changes is difficult to predict, but two trends have emerged in recent years. On one hand, some companies are now competing with energy never seen before in American markets and on the other hand, some long time rivals are joining forces to cooperate in capturing markets.

COMPANY VERSUS COMPETITOR RELATIONSHIPS

SPC is one source of ammunition in battles between companies. If company "A" can establish process capabilities before company "B", then company "A" has an edge in obtaining new contracts. However, if company B is using advanced reliability engineering to make its product last longer, then it can offer longer warranty protection to attract more customers. In either case, these companies are using the information from their quality systems to compete with each other.

At this point it should be noted that companies with SPC or other related systems have still lost large markets because they failed to link their competitive strategies to their principle of continuous improvement. Harley-Davidson is a good example of a company that did use SPC strategically. Harley-Davidson was the last surviving motorcycle company in the United States in the mid-1970s. The Japanese and other foreign companies controlled 97% of the U.S. market. The management at Harley toured the Japanese plants and discovered the better quality of Japanese bikes. While Harley had to repair almost all bikes produced before they would start, companies like Yamaha rarely found a bike that wouldn't start fresh off the assembly line. Harley quickly determined that improved quality would reduce production costs—no need for a repair area for one reason.

After a few years, Harley developed SPC and problem-solving teams and today, has doubled its market share, but not just because it manufactured an improved quality product. Sales increased because Harley aggressively marketed the new quality. Harley was represented at every motorcycle event it could find and offered free comparison rides.

Other companies have pursued new markets by bringing the quality-cost issues to the forefront of sales. One example is life cycle costs. If a customer wanted to buy a car, and had to choose between a $10,000 car, and a $12,000 car, a premature decision would be to buy the cheaper of the two. However, the customer is aware of costs not on the window sticker—maintenance, repair, fuel, and other expenses during its life cycle. If the second car gets twice the gas mileage with half of the average repair costs of the first car, the customer will see the second car as more economical in the long run.

Much of the advertising of the 1980s is playing up the total cost concept. For example, Chrysler offers seven year warranties on its products as an incentive to buyers. Other companies offer lifetime repair or return warranties on their products. If anything goes wrong with a product, just bring it back to the store and get another. All of these are examples of increases in competition resulting from the reduction of available markets.

COOPERATION

Statistical process control is one type of new technology which is causing some other changes in relationships among competitors. SPC requires training company executives in techniques of control charts and continuous improvement. Today it is not unusual to find executives of several competing companies in the same training seminar. What happened in many past SPC training sessions was that these supposed competitors found themselves working and talking together as teams. The reason for the sudden cooperation was obvious—they faced a common threat. In this case, the threat was the presence of statistics which is a subject notorious for producing a feeling of fear in the student. Many executives had little or no experience with math so they joined forces to help each other.

This situation was repeated under other conditions at many companies during the 1970s and 80s. The Japanese invasion of U.S. markets showed many companies that competing with the business across town could result in a foreign competitor blindsiding both domestic companies. Therefore, in places such as Michigan and Iowa, companies met to discuss implementation of SPC and other techniques. Suggestions were openly shared. Also during this period, large companies such as GM and Toyota began to cooperate in joint ventures. Clearly, the spirit of cooperation between suppliers and customers was spreading to competitors.

JOINT VENTURES

Joint ventures provide an excellent example of why cooperation can sometimes be the best form of competition. The reasoning behind a joint venture is that two or more companies together can create a competitive product that one of them is incapable of producing alone. In other circumstances, one company controls markets in a country which another company wants to access. Instead of competing for markets, the interested company offers technology, products or other services to the target company for rights to sell in its markets. The important point is that cooperation must benefit all participants in the project; otherwise, exploitation takes hold.

Joint ventures come in several forms; some kinds of joint ventures include:

- Two or more companies join sales forces to sell each others' product through a single distribution network;
- Several companies build a research and development center that all will share;
- One overseas company agrees to market an American product in exchange for the ability to market its product in the U.S.;
- Two companies from different countries lease production space for each company to produce its product in the other country;
- Two competing companies join forces to produce a single product that they both will sell and will split the profits.

This last type is the focus of the following case study.

CASE STUDY: THE NOVA

In 1983, Toyota and General Motors announced that they were going to use a closed assembly plant in California to jointly produce a new car model. Reaction to the proposal was mixed at the time, but by 1986 some of the advantages of this joint venture were becoming obvious.

The joint venture was formally called the New United Motor Manufacturing, Inc. (NUMMI). Both GM and Toyota agreed to contribute $100 million to the project and to share management and technology. The new product was a redesigned Corolla that used both American and Japanese parts. Vendors from both countries would share the supply contracts and the arrangement would profit both companies. For General Motors it meant having a subcompact for its dealers. GM has long been deficient in offering subcompact models. Also, GM would be able to observe and experiment with Japanese manufacturing techniques. For Toyota, the reopened plant saved building a new plant in the United States, provided access to GM's distribution network, and fought off trade restrictions.

Other advantages have surfaced from this joint venture; the main one is the change in labor relations at the plant. The old GM plant was racked with labor problems and the joint venture plant has been surprisingly free of them. The Japanese managers renegotiated the labor contract by giving the employees job security in exchange for scrapping the cumbersome work rules. In addition, problem-solving work teams encouraged employees to participate more in the running of the company. Worker morale has increased dramatically with the end result that other automotive manufacturers are implementing similar joint ventures. As a further note, many supply companies to the auto industry are now forming joint ventures with other supply companies around the world.

COOPERATION MAY BE A FORM OF SURRENDER

There is opposition to joint ventures. Some regard joint ventures with companies outside the United States as a form of surrender. From this viewpoint they see short-term gains of the venture, but warn of the long-term competitive risks. A major hazard comes from allowing overseas companies to design and operate the production system. If Americans do not learn from new production methods, their companies will become dependent upon foreign sources meaning that overseas labor will be responsible for skilled trades and technologies. The U.S. companies could become the suppliers of basic labor and distribution for foreign companies.

Joint ventures are considered as a way to dodge trade restrictions. Japan is frequently cited—if the Japanese can form dependent relationships with many U.S. firms, the United States will be less likely to impose trade restrictions for fear of hurting its own industries. As yet, these worries have not been realized. However, the success or failure of many joint ventures now under way will indicate the future of cooperation for increased competition.

SUMMARY: COOPERATION AS A COMPETITIVE STRATEGY

One reality in the new economic age that began in the 1970s is that a company can no longer compete by isolating itself from all other companies. Cooperation with suppliers, customers, professionals, and competitors is essential for increasing market shares. The number of technologies, management techniques, and customer demands outstrip the abilities of a single company. One condition related to SPC is the increased demand for product quality. Only when buyers and suppliers form a cooperative effort can strides in quality improvement be achieved. A manufacturing firm or a law office can not increase the quality of products and services if the companies supporting them are not in touch with quality needs. In the end, communications, negotiations, and personal relationships are the key to forming cooperative arrangements that benefit all.

CHAPTER 15
THE HUMAN IMPACT OF SPC

Ellen was fourteen when she left school. At age sixteen she lied about her age to get a job in a production plant. Now at age 44 she is hearing rumors about something called SPC. The word on the shop floor is that every employee will go to a class on SPC. Furthermore, there's talk that math and statistics are part of the class. No one is really sure if all of the employees will be required to use this math on the job. Ellen is terrifed. First of all, her ability to read and write is below a sixth grade level. Second, she never did well with math and finally, she has not been in a classroom in almost 30 years.

SPC is change. Change in any environment will create an impact on the people involved, but the impact does not have to be negative. By anticipating people's reactions to change, a creative manager can not only prevent adverse consequences, but can create unanticipated of benefits.

PERSONNEL AFFECTED BY SPC

In short, just about everybody involved with a company is affected by SPC. The impact expected can be divided by the type of work each group performs. For the sake of discussion, these people will be divided into five groups.

1. Hourly Workers. The hourly worker is the main source of information from the processes. Under SPC, the hourly worker is expected to take greater responsibility for the operation of the process, use more skills, and be more accountable for any action taken.
2. Unions. Since, SPC requires workers to stop and examine the performance of the process, the union is active in the renegotiation of labor agreements. Issues such as piece rates and the ability of operators to call for help beyond their direct supervisors are hotly discussed issues.

3. Management and Office Personnel. Both management and office personnel become support groups to the SPC effort. Instead of providing all directives about the production process, these people must now learn how to obtain useful feedback from production workers. Authority is replaced by a system of working partnerships among departments.

4. Suppliers and Subcontractors. SPC almost always requires extra work from the supply companies and most of the subcontractors. Since quality from the source is one principle of SPC, these companies must form working relationships and information exchange programs.

5. Customers. Customers derive many benefits from an SPC system. There are the obvious ones of higher quality and lower costs and less conspicuous ones, such as more attention being paid to their opinions.

The issues related to each group are the same issues confronted whenever a new technology is introduced. Change in method of management also requires change in the way people interact. While such changes may promote conflicts among people and departments, they can also encourage a much needed spirit of cooperation. The result a company experiences will be directly related to how well management will plan and "sell" the new system. It is strongly recommended that a company wishing to implement SPC talk to other companies that have gone through a similar experience to make them aware of many minor issues that could create unexpected problems.

SPECIFIC PROBLEMS

For all employees within a company implementing SPC, there will be a number of specific issues which management must address to avoid problems if they are left unresolved. Some of these issues are:

1. Illiteracy
2. "Math blindness" and math anxiety
3. Language as a barrier
4. Training
5. Retraining
6. Decentralized authority

Illiteracy

There is an ongoing controversy about the exact level of illiteracy in this country. Current estimates say that between one-in-ten to one-in-eight Americans do not have the skills needed to complete a job application form. Some of these people left school; others never got the attention they needed while in school. Whatever the reason, the illiteracy rate in the United States is high enough that it must be considered before any SPC system is implemented.

A good personnel department will have some idea of the people who are weak in reading and writing skills. However, companies have to approach the problem with extreme tact because of the stigma associated with illiteracy. Many illiterates have devised complex methods of avoiding detection. Others are protected by their co-workers. The best method for management is to introduce information gradually about programs available for increasing reading and writing skills. Local school systems will have information about the programs in the area. The message must be clearly communicated that it is never too late to learn how to read and write properly.

Care should be taken to find the people with weaker reading and writing skills. It may be difficult to separate them from their work team. In most cases, the instructor for the SPC course should be notified and special adaptations made. One recommended addition to training sessions is to allow time for students to meet privately with the instructor which helps the instructor provide individual attention to special needs.

Once SPC becomes part of the work function, supervisors should be trained to watch for signs of someone having difficulty with the charts. Indicators include incomplete charts, missing information, improperly placed plotting points, or unreadable writing. In such cases, the supervisor should be instrumental in creating adaptations to help people feel more at ease with the SPC system.

In a welding facility outside of Louisville, the weld inspector used to inspect welds by marking flaws with a white grease pen. The company implemented an SPC system that would require him to count and write down the type of defects detected. Unfortunately, his supervisor knew that this young man left school in the fifth grade to help with the family farm. Numbers and writing were not his strong suit, so the supervisor supplied him with a picture of the assembly to inspect. The inspector would then make an "X" on the picture where each defect was found. This is the same defect map used for the c-bar chart in Chapter 3. At the end of each hour, the supervisor would count up the number of defects for the chart.

"Math Blindness" and Math Anxiety

Illiteracy is a major handicap to implementing SPC, but is minor in comparison to two conditions that are far more prevalent. These disorders affect employees regardless of their level of education or position in the company. They are called "math blindness" and "math anxiety." "Math anxiety" is the common fear many people have of mathematics. The fear ranges from insecurity with using numbers to the inability to calculate under any type of pressure. Most people with math anxiety will take steps to avoid making calculations, especially when it involves learning new methods. Estimates of the extent of problem are unreliable, but it is usually safe to assume that over half of any work force has this condition.

The other condition is called "math blindness" because it represents a high level of math anxiety from a specific cause. The "math blind" individual is a person who never learned mathematics. The typical case is the student who was actively discouraged from learning math in any form. The largest group of the "math blind" are women. These women lived during a time when math was considered unnecessary for their education. As a result, they cannot recognize equations.

What makes these problems different from those related to illiteracy is that many of these people are well-educated and sometimes hold management level jobs. Two steps in training can help reduce the effects of these problems. The first is to make it clear to the participants of training that the need to use math in SPC will be kept to a minimum for the people filling in charts. To further help with this message, some companies offer a class in basic math skills before training begins for people wanting to sharpen their abilities. Many times classes are also attended by people who enjoyed math in school, but haven't had to work math problems for many years.

The second step is to give every employee a simple, four function calculator and the SPC training should include a brief session on how to use it. It should never be assumed that since these calculators are readily available, they are widely used. It has been the author's experience that even in training sessions for top management, many of the participants are unfamiliar with calculators. Calculators are particularly effective because employees can take them home and practice in private.

Language as a Barrier

The English language is tricky to learn and full of ironies and idiosyncrasies. For people who learned English as a second language, this can quickly form a barrier to learning. When possible, SPC methods should be taught in the language in which people feel most comfortable. In the United States, the most common exception to English will be Spanish. A bilingual instructor is always an asset. Caution is recommended, since all languages differ by location. The Spanish spoken in Arizona will be different from that in Spain. French in Paris is very different from that in New Orleans.

One language that must be used in some classes is sign language. The deaf need sign language, a clear speaking instructor, and some preparation before class. For example, the instructor, signer, and deaf students should meet before training begins to lay some ground rules. Most of these rules will deal with how to sign some of the statistical phrases such as "statistical process control", "sigma", and other related terms that are not part of daily language. Other agreements have to be reached on what parts of the course will be written. During calculations, it is sometimes desirable for the facts of the example be noted at the top of the page. Otherwise, the participant may be overwhelmed by incoming information from the signer, and may not follow the steps during calculation.

Training America

As can be seen in the above discussion, any training effort will involve a combination of personalities and skills. Conflicts, fears, and resistance are common during the implementation of SPC. Nevertheless, the training phase is critical for eliminating most of the problems of implementation. The best step that can be taken at this point is to introduce the teamwork concept. Participants with previous SPC experience, or strong skills related to SPC should be encouraged to help their fellow employees. Production teams are usually well-known for protecting their own. Such feelings should be encouraged and management should take every opportunity to help foster team efforts.

One way to encourage teamwork is to involve each class or work group in planning the SPC system for their area. The instructor, management, and the team could meet to decide the finer points about which charts to use in their areas and which characteristics to measure. A member of top management should be in each class to answer concerns or questions about the policies related to SPC. Finally, each participant in a class should have a part of the SPC system available at the completion of training. For example, a control chart should be part of the job by the end of training. The participant should know that perfect use of the chart is not expected immediately. In fact, many companies have the employees try the chart for a week or so, and then the instructor goes to their work stations to answer questions.

Retraining the Employees

For most of people who will become part of the SPC system, the issue is training, i.e., the addition of a few skills to allow them to participate in the system. Their job function remains about the same. However, for a small group of people, the issue is retraining. In these cases, a person is ending one type of job and is being retrained for an entirely new position. Examples include the housewife being trained to become a machine operator or the layed-off machine operator being trained to become an SPC Coordinator.

Retraining is a separate issue because of the long training normally involved and the dramatic change in the job function the participant experiences. The example of the machine operator becoming an SPC coordinator is a good example. Such a person one day is worrying about RPM's, stock shortages and blueprint specifications and the next day is learning how to operate a computer and is calculating standard deviations. The extended period of adjustment must be designed into an extended schedule of training. Part of this training should be working side-by-side with someone already performing the function, as a form of internship.

Decentralized Authority

Other human impacts will occur after training. The most obvious of these will be the decentralization of authority. The SPC system requires that machine operators monitor and correct processes on their own initiative. Some plants go as far as to give individual operators the ability to shut down a production line to fix their problems. In many cases, this is a dramatic change from an older system of management which dictated that only management could make a decision. The role of the operator was to perform the job without questions.

Work teams, problem-solving efforts, control charts, and participatory management will dramatically change the landscape of power and politics within a company. The group most affected by such a change is management. Line supervisors, foremen, and managers will tend to resist a system that gives decision-making powers to all employees because of their concern about decentralization of authority. The old system is no doubt familiar to the management and rules of company politics have been established. SPC tends to wipe away many of these internal understandings so that management must develop a new perspective.

Management must realize that SPC represents an even bigger change. The system is designed so everyone can contribute to the good of the company. The survival and growth of the company is where their future is in this new age of economics. There no longer is room for company politics and the change is extremely difficult. A wise chief executive will create an environment that encourages activities that support SPC.

At a major manufacturing firm in Michigan, the CEO met monthly with all of the employees. During one meeting the CEO explained how reaction plans for SPC worked along with the idea that deviations in the process should be brought to the attention of supervisors. The CEO put a number on the overhead projector and said,

"Now if you don't feel that your supervisor cares about correcting what you discover, this is my home phone number. Call me any time and tell me about it."

This type of bold action can be clearly interpreted by middle managers.

CHANGE IN LABOR REQUIREMENTS AND AGREEMENTS

An impact on labor relations is almost inevitable in an SPC system. It is recommended that labor negotiations include the SPC implementation plan as soon as possible. Filling in of charts and the reaction to out of statistical control points are just two conditions that will change existing job functions. Under an agreement where each operator is paid according to the number of parts produced each hour, there is no encouragement for an operator to stop and chart. The charting time reduces take-home pay. Existing labor agreements have to be redesigned to cover piece work employees.

Previously, a worker may have been trained that problems should be ignored or reported verbally to the supervisor. The SPC system will require reaction and documentation from the worker. If the supervisor does not support the change, then no change will occur because the worker does not want to be in conflict with the supervisor.

Labor negotiations and flexibility in personnel practices will help to solve these types of problems but these steps will only be helpful when the company has a clear picture of the system it wishes to create. It is advisable to have a good strategic plan and open discussions about SPC for a smooth implementation.

SOLUTION: HUMAN RELATIONS AS A SOURCE OF QUALITY AND COMPETITION

A popular argument states that 85% of problems experienced in a company's production process are caused by management. In reality, all problems in production are the responsibility of management. The former argument was based on the idea that 15% of the problems were caused by the line workers making mistakes or were difficulties line workers could correct. Recent developments in manufacturing have made this argument obsolete. It is now being shown that new companies have studied the problems of older production lines and have designed the causes out of the system. In other words, a mistake by a line worker is the fault of a management group that did not take the time to conduct proper training or to design methods of quality assurance into the process.

The solution to human problems in a production system, including the SPC techniques, is leadership. A company that has strong leadership can confront and eliminate the cause of any problem. In military science this is known as the principle that any well-organized and inspired group can overwhelm any defensive position. A good leader within an SPC environment is one who will walk around the plant and help support the work teams. A good leader will actively develop the skills of the work force. Training, procedure reviews, and work groups are all part of the continuous improvement of the work force. A successful leader will also clearly communicate the goals of the company and will engender a belief in those goals.

Charisma is not the only possible way to instill a common conscience about improvement. Consistent integrity will have the same effect because people will respect such leadership. When people believe and trust in a company, there is very little they cannot accomplish as a team. Tom Peters' works are filled with examples of companies that have this type of leadership which came from the top manager as the most inspired believer in the company's goals. In the case of SPC, the best leader is one who is dedicated to continuous improvement.

SUMMARY

In Alvin Toffler's book, *The Third Wave*, he described how America is changing from a manufacturing economy to a service economy and predicted that information would become the nation's number one product. SPC represents the crest of the third wave. The SPC system requires that every product must be accompanied by information. This is the core of the impact of humans working with SPC. Each employee now needs the ability to read, write, and calculate to excel in the work place.

To create a work force and management team that can effectively use SPC, a company must embark on a massive program of training and retraining. To prepare people for it, they must be screened so that weak skills will be improved. Clear communication about the training and the SPC system must be maintained. Each person should be given the chance to participate in the system. Finally, each success must be advertised so the benefits achieved are felt by everyone.

APPENDIX A
ANSWERS TO QUESTIONS IN CHAPTERS

PROBLEMS FOR CHAPTER 2

1. Complete the following X-bar/R chart.

Time	3	4	5	6	7	8	9	10	11	12	1	2	3
	66	63	65	66	65	59	65	60	65	67	60	59	65
	65	63	67	63	60	60	64	63	61	67	65	59	65
Data	62	61	64	63	60	61	66	65	61	61	65	61	61
	68	59	61	59	61	61	64	61	66	62	64	63	63
	67	65	65	64	62	61	64	61	65	63	64	60	62
Total	328	311	322										
Average	65.6	62.2	64.4										
Range	6	6	6										

2. For the following sets of data, calculate the mean, standard deviation, Cr, Cp, and Cpk ratios.

a)

1	8	1	2	6
0	4	2	5	8
5	5	3	3	2
6	6	4	6	4
3	1	5	3	1

Specifications: 3 ± 5

b)

13	14	18	13	14
16	11	14	13	13
14	11	16	9	14
13	18	14	13	9
13	13	14	11	14

Specifications: 13 ± 2

c)   .57   .44   .54   .51   .38   .61   Specifications: 0.50 ± 0.25
     .54   .56   .44   .57   .44   .56
     .57   .58   .53   .46   .62   .32
     .56   .41   .44   .48   .48   .49
     .41   .52   .35   .42   .64   .40

3. Using histograms, check the above sets of data for a normal distribution. Which sets appear to have a normal distribution?

PROBLEMS FOR CHAPTER 3

1. Calculate the control limits for a p-bar chart using the following information.

        sample size = 100 pcs.

p =   .10   .07   .10   .10   .08
      .09   .08   .07   .08   .08
      .08   .08   .11   .09   .09
      .09   .08   .09   .09   .08
      .09   .08   .08   .10   .09

2. Calculate the control limits for an np-bar chart using the following information.

        sample size = 120 pcs.

np =   21   12   25   14   18
       20   21   21   22   13
       37   14   26   26   23
       22   30   9   17   18
       12   15   17   12   24

3. Calculate the control limits for a c-bar chart using the following information.

sample size = 1

c =   2   1   6   2   5
      5   0   7   2   8
      3   5   3   6   5
      5   3   4   5   6
      1   2   1   6   5

4. Calculate the control limits for a u-bar chart using the following information.

sample size	u	sample size	u	sample size	u
9	14	7	13	7	17
8	11	7	41	8	33
7	18	9	25	8	21
8	13	6	37	9	26
8	21	10	40	8	28
6	31	8	19	7	36
7	11	7	41	8	29
7	12	9	21	7	14

PROBLEMS FOR CHAPTER 5

1. Identify the first point out of statistical control on the following control charts.

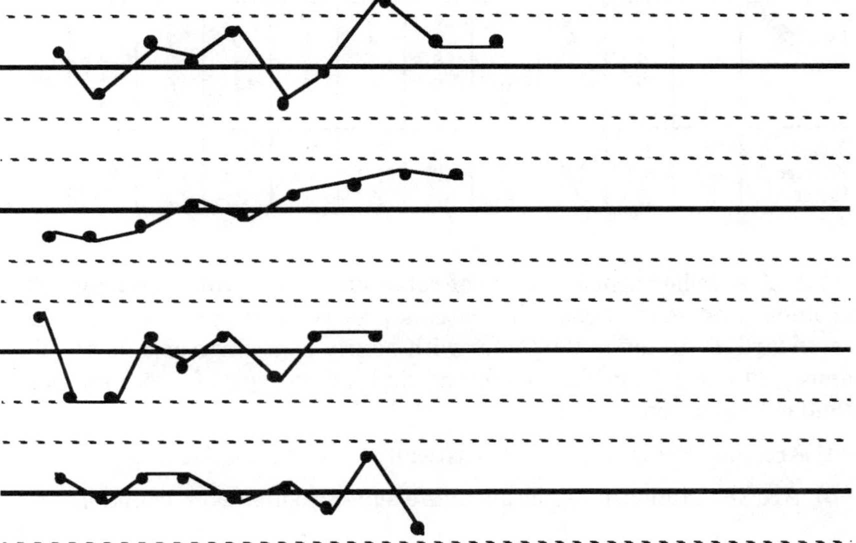

2. A process makes an average part 6 inches long. The standard deviation from a sample of 30 pieces is 0.05 inches. The specifications call for a length of 6 ± 0.20 inches. Calculate the Cpk index for this process. Does the process show the potential for capability?

3. The customer for the above part wants 10,000 built to a tolerance of 6 ± 0.10 inches. It will cost $0.05 to sort each part out of specification. The company's profit margin on this project is $500. Should it proceed with production?

PROBLEMS FOR CHAPTER 6

1. Find as many errors as you can from the control chart below.

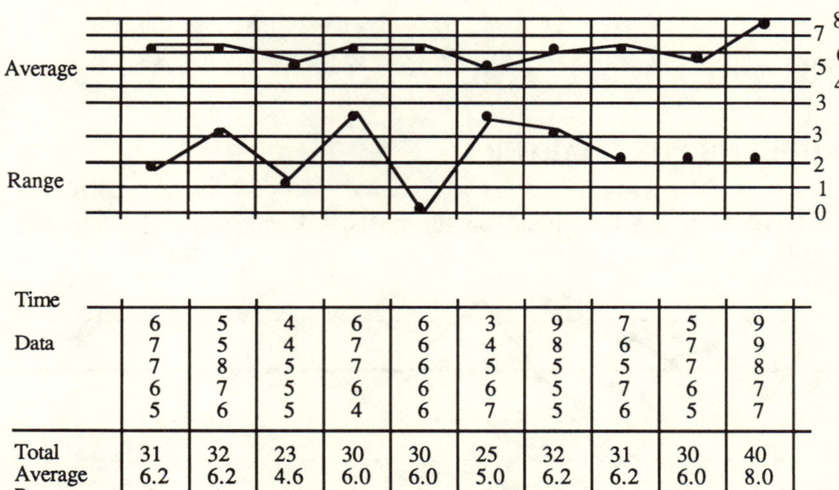

Time										
Data	6	5	4	6	6	3	9	7	5	9
	7	5	4	7	6	4	8	6	7	9
	7	8	5	7	6	5	5	5	7	8
	6	7	5	6	6	6	5	7	6	7
	5	6	5	4	6	7	5	6	5	7
Total	31	32	23	30	30	25	32	31	30	40
Average	6.2	6.2	4.6	6.0	6.0	5.0	6.2	6.2	6.0	8.0
Range	2	3	1	4	0	4	4	2	2	2

2. A supplier sends a batch of retaining pins to your company. The accompanying control chart indicates a process average of 3.01 mm and a standard deviation of 0.02 mm, with hourly samplings of 5 parts. You sample 30 parts from this batch and find an average of 2.99 mm and a standard deviation of 0.015 mm.

a) Are the means different statistically? (assume 95% confidence)

b) Are the standard deviations different? (assume 95% confidence)

PROBLEMS FOR CHAPTER 7

1. Using the following raw data collected from the audit of paperwork accuracy, fill in the following defect grid.

Fred—3 spelling, 3 math, 1 grammar, 1 spacing, 1 no signature,
    1 other
Sally—4 spelling, 3 grammar
Bob—1 spelling, 1 spacing
Jan—2 spelling, 2 math
Doreen—11 spelling, 3 grammar
Tom—2 spelling

People

Problems:	Fred	Sally	Bob	Jan	Doreen	Tom	Total
Spelling							
Math							
Grammar							
Spacing							
Missing Sign.							
Wrong Form							
Other							
Total							

2. Suggest possible problem-solving steps for the above situation.

3. Assume that 3 pages were examined for each person involved. Create control limits for a c-bar chart using the information in problem one.

PROBLEMS FOR CHAPTER 9

1. A tool loses 0.001 inches of cutting size for every 5,000 strokes. Current control charts place the process average at ten inches. How many strokes before the specification of 10 ± 0.005 inches is exceeded?

2. Using the above information and assuming a standard deviation of 0.0005 inches, how long until the ±3 sigma limit touches the specification limits?

3. Each of three parts wired in series has a 0.995 chance of success. When operating as a single component, what is the chance for failure?

4. What is the chance for failure in the following curcuit?

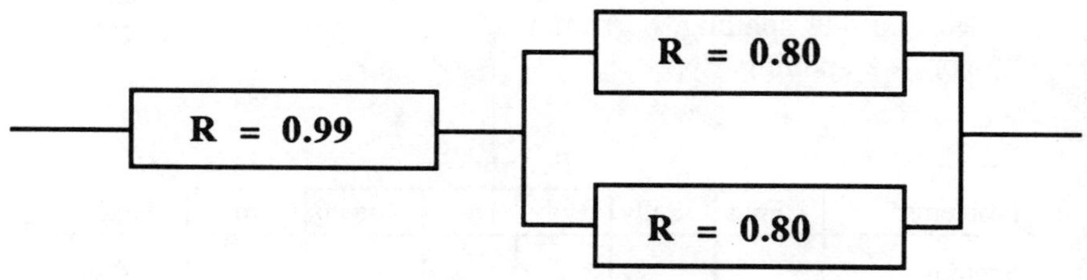

5. Failure rate calculation. A component fails an average of 15 times in 20 hours of operation. What is the failure rate per hour?

6. Plot the following failure rates over time. What type of curve is formed?

Time	Failure Rate
0-10 hrs	3.1
10-20	2.5
20-30	1.1
30-40	0.9
40-50	1.2
50-60	2.1
60-70	3.6

PROBLEMS FOR CHAPTER 10

1. Calculate the loss function for the following:

   a) Specification = 15 ± 3; Average of 14.5; Cost to scrap = $1.25
   b) Specification = 125v ± 5; Average of 124v; Standard Deviation of 2.2v; Cost to scrap = $2.50
   c) Specification = 10.5 ± 0.25 mm; Average of 10.5 mm; Standard Deviation of 0.025 mm; Cost to scrap = $100

2. Calculate the reduction in loss from a process that reduces the variation in parts from 0.50 to 0.30. Assume an average of 15, a cost to scrap of $30.00 and a specification of 15 ± 1.0.

3. Using the z-value formula, calculate the reduction in scrap using the above information.

PROBLEMS FOR CHAPTER 11

1. The management of a small manufacturing company is provided data for new methods of producing three existing products. Which methods should be seen as improvements? What do you choose as your criteria?

	Old Method	**New Method**
Product A	Cpk = 1.00 Scrap = 5% Capacity = 5000 per hour Mean Time Between Failures = 500 hrs.	Cpk = 1.20 Scrap = 2% Capacity = 4800 per hour Mean Time Between Failures = 1100 hrs.
Product B	Cpk = 1.33 Scrap = 5% Capacity = 1000 per day Loss Function = $1.25	Cpk = 1.33 Scrap = 5% Capacity = 1200 per day Loss Function = $1.30
Product C	Loss Function = $12.00 Capacity = 10,000 per week Sorting Cost = $1,000 per day Profit Margin per day = $5,000	Loss Function = $10.00 Capacity = 8,000 per week Sorting Costs = $0 Profit Margin per day = $500

2. A process improvement will cost $5,000 however, it will decrease scrap from 5% to 3.5%. Each scrap piece will cost $3.00 to sort. 10,000 pieces are made each month. How many months until a return on the investment will be realized?

ANSWERS TO CHAPTER PROBLEMS

Chapter 2:

1. The completed control chart.

Time	3	4	5	6	7	8	9	10	11	12	1	2	3
Data	66	63	65	66	65	59	65	60	65	67	60	59	65
	65	63	67	63	60	60	64	63	61	67	65	59	65
	62	61	64	63	60	61	66	65	61	61	65	61	61
	68	59	61	59	61	61	64	61	66	62	64	63	63
	67	65	65	64	62	61	64	61	65	63	64	60	62
Total	328	311	322	315	308	302	323	310	318	320	318	302	316
Average	65.6	62.2	64.4	63	61.6	60.4	64.6	62	63.6	64	63.6	60.4	63.2
Range	6	6	6	7	5	2	2	5	5	6	5	4	4

2. a) Average = 3.76
   Standard Deviation = 2.2226
   Cr = 1.33
   Cp = 0.75
   Cpk = 0.64
   b) Average = 13.4
   Standard Deviation = 2.1985
   Cr = 3.30
   Cp = 0.30
   Cpk = 0.24
   c) Average = 0.4947
   Standard Deviation = 0.08316
   Cr = 1.00
   Cp = 1.00
   Cpk = 0.98

Chapter 3:

1. p-bar = 0.0868
   UCL = 0.17
   LCL = 0.002

2. np-bar = 19.56
   UCL = 31.70
   LCL = 7.42

3. c-bar = 3.92
   UCL = 9.86
   LCL = 0

4. u-bar = 3.115
   UCL = 4.20
   LCL = 2.03

Chapter 5:

1. The circled points are the first to be out-of-statistical control.

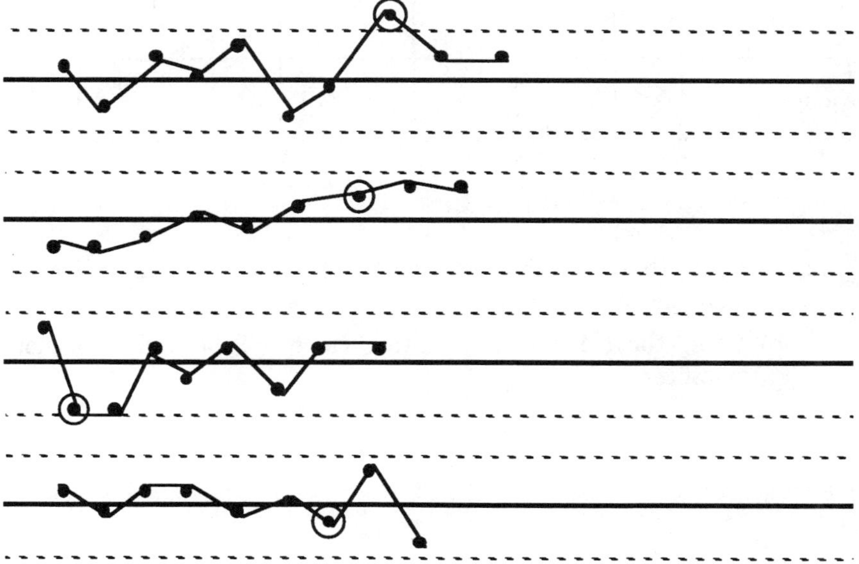

2. The Cpk is calculated to be 1.33 therefore, this process is capable.

3. No. Using the Z-value formula we find that there will be 2.28% scrap at each extreme of the distribution. This is a total of 4.56% scrap. Multiplied into the 10,000 part run estimates a cost of $2,280.

Chapter 6:

1. Some of the errors on the control chart are noted below.

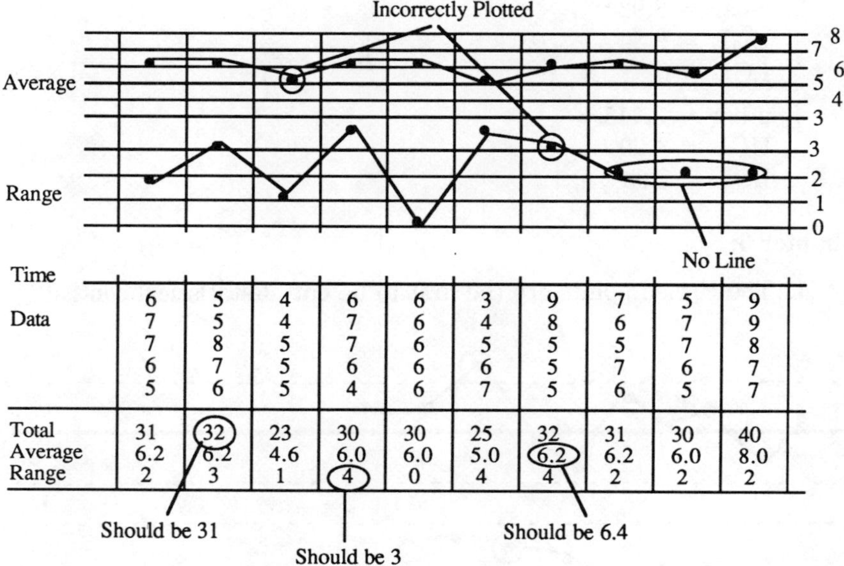

	1	2	3	4	5	6	7	8	9	10
Data	6	5	4	6	6	3	9	7	5	9
	7	5	4	7	6	4	8	6	7	9
	7	8	5	7	6	5	5	5	7	8
	6	7	5	6	6	6	5	7	6	7
	5	6	5	4	6	7	5	6	5	7
Total	31	32	23	30	30	25	32	31	30	40
Average	6.2	6.2	4.6	6.0	6.0	5.0	6.2	6.2	6.0	8.0
Range	2	3	1	4	0	4	4	2	2	2

2. a) Using the t-test the resulting t-value is 2.138. This indicates a significant difference between means.

   b) Using the F-ratio the resulting value is 2.149, which is not significant.

Chapter 7:

1. The resulting grid of defects is

People

Problems:	Fred	Sally	Bob	Jan	Doreen	Tom	Total
Spelling	3	4	1	2	11	2	23
Math	3			2			5
Grammar	1	3			3		7
Spacing	1		1				2
Missing Sign.	1						1
Wrong Form							0
Other	1						1
Total	10	7	2	4	14	2	

2. Since spelling is the largest problem, then initial actions should focus on improving proofreading and spelling skills. Grammar and math would be the next problems to be addressed.

3. The c-bar for a single page would be 13, therefore the control limits would be: UCL = 23.8 and LCL = 2.2.

Chapter 9:

1. It will take 25,000 strokes.

2. The three standard deviations reduce the distance to the specification to 0.0035 inches therefore, only 17,500 strokes are needed before repair.

3. 1.5%

4. 5%

5. 0.75 per hour.

6. A "bathtub" curve.

Chapter 10:

1. a) $0.035
   b) $0.584
   c) $1.00

2. The loss-function before is $0.75 but, after it is $0.27. This results in a reduction of $0.48.

3. The old system would create about 4.56% total scrap, the new system would create only 0.086% scrap.

Chapter 11:

1. The answers to all three situations would depend heavily upon the criteria of "success" management has stated in the company's strategic plan.

   a) The new method generates less scrap and production. However, if the number of successfully made parts are considered, the production levels are almost the same. The new method is better.

   b) A close decision. The new method raises production, but also creates higher internal costs. More economic data would be needed to make a clear-cut decision.

   c) This decision is either hard or easy, based on the criteria the company has adopted. If profits are the major criteria, then the old method would be chosen. However, the new method is superior for reducing internal costs.

2. 11.1 months until pay back.

APPENDIX B
GAGE REPEATABILITY AND REPRODUCIBILITY STUDIES

Chapter 4 noted that to ensure the integrity of the data gathered in an SPC system, all measurement gages must be under strict control. Part of the control is to physically track the location and use of each measurement gage in a company. Part of tracking is to schedule each gage for calibration at least once a year. For delicate or critically important gages, such as those in the labs, calibration should be done more frequently. The purpose of calibration is to determine gage accuracy. However, all measurement gages must be submitted to a repeatability and reproducibility study.

GAGE ACCURACY AND MEASUREMENT ERROR

Gage accuracy is the difference between the observed average measurement and the true size of the part being measured. An example of this is the calibration of a vernier caliper. The caliper is sent to a certified measurement testing lab for evaluation. A set of measuring blocks, certified as being a specific size, is measured by the caliper. The test may use a one inch standard block where one caliper is used to measure the block several times. The average of these measurements is called the observed average. The true size of the block is certified at one inch. If the caliper finds an average measurement of 1.005 inches, then the gage is said to be within 0.005 of an inch of accuracy.

It is up to the company and its customers to determine the limits of allowable error. In the above case, the gage is being used to measure parts with tolerances measured in the hundreths of an inch. The relatively small amount of gage accuracy error is not seen as significant. If the accuracy were off by greater amounts however, the company might instruct the testing lab to reset or repair the caliper. If that proves im-

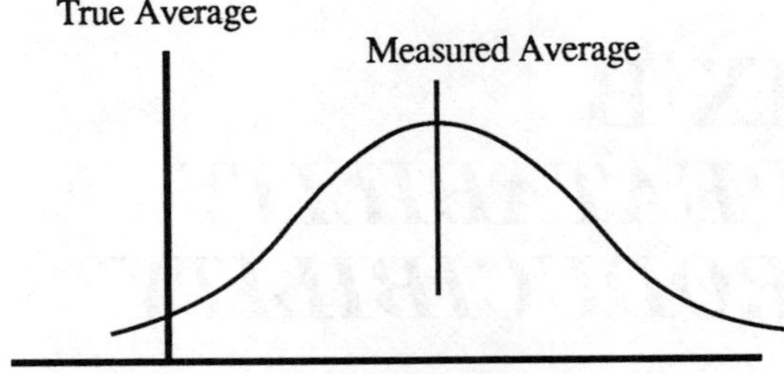

Gage Accuracy -
The difference between
the true average size of
parts and the measured
average.

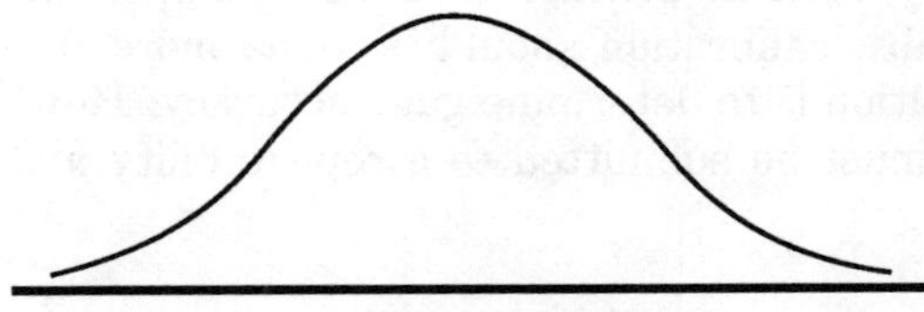

Repeatability - the spread of
repeated readings from a
single gage measuring the
same part repeatedly.

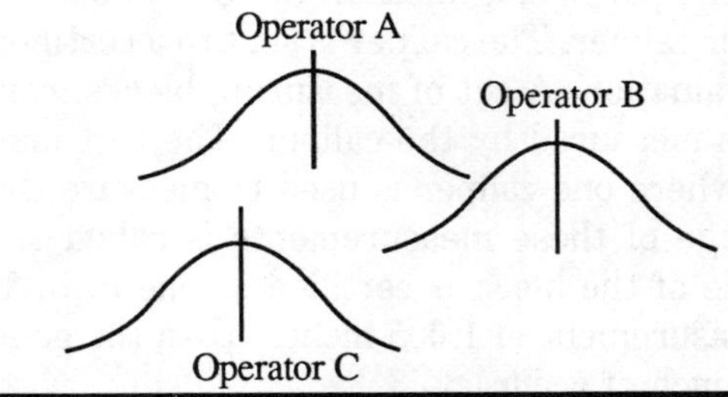

Reproducibility - the difference
between the averages obtained
by each operator.

Figure B-1 The concepts of gage error, repeatability, and
reproducibility.

possible, then the gage is removed from its location and a more accurate one is purchased. These costly precautions are necessary. If the accuracy of a gage is off significantly to begin with, then it will only cause harm during data collection by reporting erroneous measurements.

REPEATABILITY/REPRODUCIBILITY

Once gage accuracy has been demonstrated, the next question is the amount of repeatability and reproducibility error present in the gage. While gage accuracy looks at the average performance of a gage in standard tests, a repeatability and reproducibility study will examine both the average and variation of measurements on actual parts. The gage repeatability and reproducibility study is divided into three parts. The first part is to evaluate the amount of repeatability error present in a specific gage. *Repeatability* is the variation observed when one operator measures the same characteristic on the same parts. The caliper from the above example has been calibrated and brought back onto the shop floor and is used to measure the lengths of small threaded rods. As a test of repeatability, the operator from the first shift is tested. This operator is handed a set of, perhaps, five threaded rods and is told to measure their length. As each measurement is taken, the tester writes down the results. Then the operator is given the same five parts to measure again for length. The measurement is repeated a third time. The idea is to see how much variation occurs between measurements on the same part. Bias is avoided if the operator is told that fifteen parts were measured instead of five parts three times. This can be done by presenting one part at a time from a box below a table.

The repeatability is estimated by taking the range of each part measured. The five resulting ranges are then averaged. Later this average range is combined with the results from other operators of the gage. This grand average range is used to calculate the amount of fixture variation present while measuring a specific part. The amount of repeatability error is estimated by dividing the fixture variation by the tolerance range for the dimension under study.

Gage Reproducibility is obtained by repeating the above test with up to three operators of the gage. Each operator will calculate an average reading from the fifteen measurements made. The difference in the averages obtained by each operator is called the appraiser variation and indicates how well different operators agree on the average size of the same set of parts. It is used to calculate the reproducibility error and is

The calculation box for each operator.

Operator A

Sample Number	First Trial	Second Trial	Third Trial	Range
1				
2				
3				
4				
5				
Sum				

$$\overline{R}_A = \boxed{}$$

$$\overline{X} = \boxed{}$$

The difference between the three readings on the same part forms the range.

Ranges are totaled together.

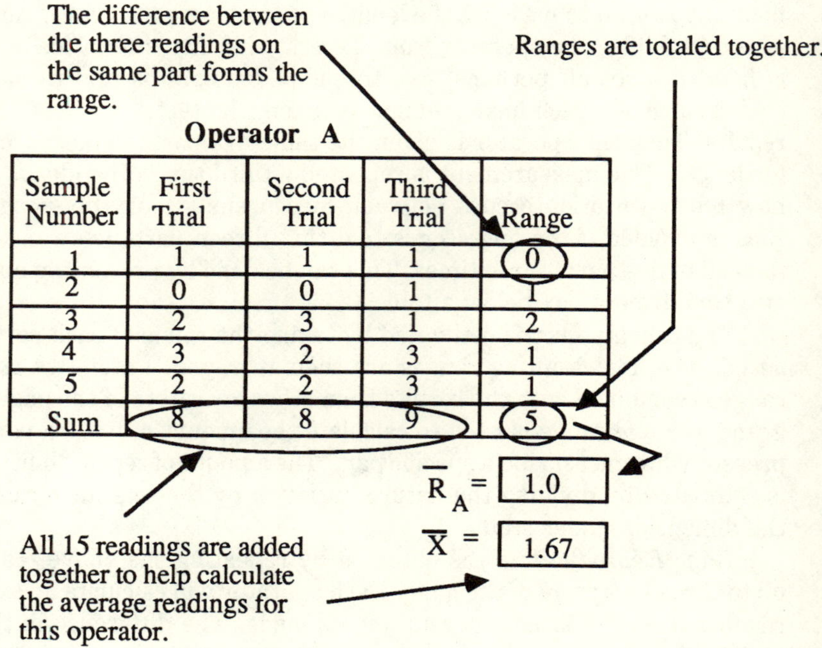

Operator A

Sample Number	First Trial	Second Trial	Third Trial	Range
1	1	1	1	0
2	0	0	1	1
3	2	3	1	2
4	3	2	3	1
5	2	2	3	1
Sum	8	8	9	5

$$\overline{R}_A = \boxed{1.0}$$

$$\overline{X} = \boxed{1.67}$$

All 15 readings are added together to help calculate the average readings for this operator.

Figure B-2 How the R&R paper is filled out.

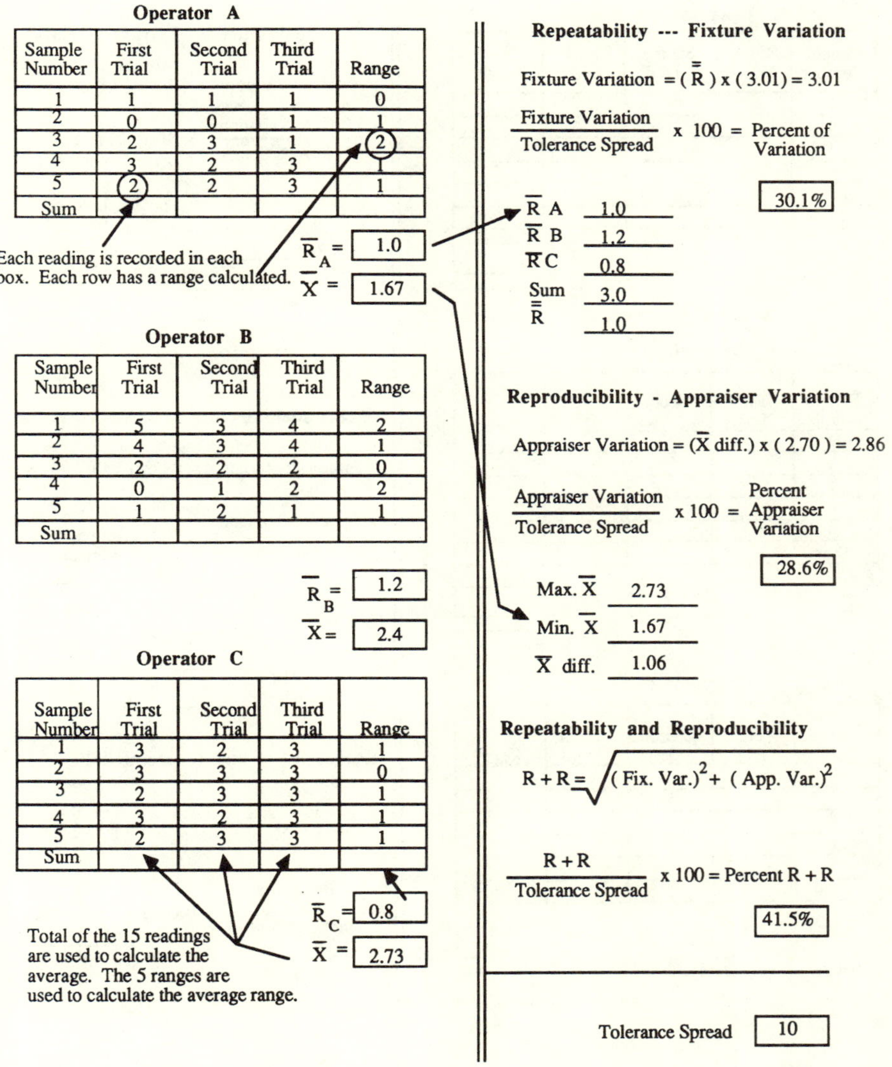

Figure B-3 Gage repeatability and reproducibility calculations.

Operator A

Sample Number	First Trial	Second Trial	Third Trial	Range
1				
2				
3				
4				
5				
Sum				

$\overline{R}_A =$ ☐

$\overline{X} =$ ☐

Operator B

Sample Number	First Trial	Second Trial	Third Trial	Range
1				
2				
3				
4				
5				
Sum				

$\overline{R}_B =$ ☐

$\overline{X} =$ ☐

Operator C

Sample Number	First Trial	Second Trial	Third Trial	Range
1				
2				
3				
4				
5				
Sum				

$\overline{R}_C =$ ☐

$\overline{X} =$ ☐

Repeatability --- Fixture Variation

Fixture Variation $= (\overline{\overline{R}}) \times (3.01)$

$\dfrac{\text{Fixture Variation}}{\text{Tolerance Spread}} \times 100 =$ Percent of Variation ☐

$\overline{R}$ A ______

$\overline{R}$ B ______

R C ______

Sum ______

$\overline{\overline{R}}$ ______

Reproducibility - Appraiser Variation

Appraiser Variation $= (\overline{X}\ \text{diff.}) \times (2.70)$

$\dfrac{\text{Appraiser Variation}}{\text{Tolerance Spread}} \times 100 =$ Percent Appraiser Variation ☐

Max. $\overline{X}$ ______

Min. $\overline{X}$ ______

$\overline{X}$ diff. ______

Repeatability and Reproducibility

$R + R = \sqrt{(\text{Fix. Var.})^2 + (\text{App. Var.})^2}$

$\dfrac{R + R}{\text{Tolerance Spread}} \times 100 =$ Percent R + R ☐

Tolerance Spread ☐

Figure B-4 Gage repeatability and reproducibility calculations.

expressed as a percentage of the tolerance range of the dimension under study.

The third part of the test involves averaging the two types of error to create the total repeatability and reproducibility error. To do this, the fixture variation and the appraiser variation are squared and added together. The square root of the results is divided into the tolerance range. The result is a percentage of gage repeatability and reproducibility error.

The standards used to evaluate repeatability, reproducibility and total repeatability and reproducibility are:

1. Zero to 10%—Gage is deemed acceptable.
2. 10% to 30%—Generally not acceptable unless gage is particularly complex. Every effort should be made to correct the accuracy of the gage.
3. Over 30%—Not acceptable.

Usually, the cause of problems with gages can be determined from the results of the gage repeatability and reproducibility study. If repeatability error alone is high, then the gage is probably in need of repair or lacks the precision required for the job. When reproducibility error alone is high, operators need more training in the proper use of the gage or a clear gage use procedure must be written and enforced. When total repeatability and reproducibility is high and each type of error is about equal, it could be a combination of effects, or it could be the gage is just not suited for the job at hand.

A STORY OF REPEATABILITY AND REPRODUCIBILITY

A story can better illustrate the use of the repeatability and reproducibility study. A plastic blow molding company measured the volume of the bottles it made by first filling a bottle with distilled water and measuring the tare weight. The weight of distilled water is directly proportional to volume. The procedure was to take an empty bottle, place it on an electronic scale, and zero the scale. The bottle was filled to the brim with water and weighed again. The weight was recorded.

One day the engineers came around to do a repeatability and a reproducibility study on this method of measuring volume. The volume tolerance was a range of ten cubic centimeters. Each operator was given five bottles to measure and each bottle was measured three times. An operator from each of the three shifts was required to perform this task.

The engineers used the resulting measurements to discover a high repeatability and reproducibility error of 58%.

Further investigations discovered the cause of the problem. First, there was a large problem with repeatability. Careful examination of the original data indicated that the measurements increased with each repetition of measurement. This was easy to understand once everyone realized the bottles were stretching under the weight of the water. To correct this problem, the engineers enlisted the aid of the industrial nurse and used her medical supplies to encase five bottles in plaster casts to prevent the bottles from stretching. The repeatability and reproducibility study was repeated. This time the repeatability error dropped to only 8%, but the reproducibility error was still too high.

An inquiry was launched to see if the operators were following procedures correctly. Once all three operators gathered for the interview, it was obvious what the problem was. The first and second shift operators were both men over six feet in height. The third shift operator was a woman under five feet in height. The engineers remembered that the scale was located on a high table. In fact, the surface of the scale was almost five feet off the ground. The tall men could fill the bottle while it was on the scale. The woman had to fill the bottle on the table and then lift it up to the scale. Thus, the bottom of the bottle would sag under the weight of the water when lifted. More water had to be added to bring the level to the brim once the bottle was on the scale which created consistently higher readings on the third shift. The supervisor solved the problem by sawing off part of the legs from the table holding the scale. At last, the repeatability and reproducibility error dropped to a more acceptable 9.5%.

Appendix C
Critical Values of t-values

Degrees of Freedom	5% level of Significance	1% level of Significance
1	12.71	63.66
2	4.30	9.92
3	3.18	5.84
4	2.78	4.60
5	2.57	4.03
6	2.45	3.71
7	2.36	3.50
8	2.31	3.36
9	2.26	3.25
10	2.23	3.17
11	2.20	3.11
12	2.18	3.05
13	2.16	3.01
14	2.15	2.98
15	2.13	2.95
20	2.09	2.84
25	2.06	2.79
30	2.04	2.75
50	2.01	2.68
∞	1.96	2.58

Appendix D
Critical Values for Chi-Square

(5% level of Significance)

Degrees of Freedom	Critical Value	Degrees of Freedom	Critical Value
1	0.004	1	3.84
2	0.103	2	5.99
3	0.352	3	7.82
4	0.711	4	9.49
5	1.145	5	11.07
6	1.635	6	12.59
7	2.167	7	14.07
8	2.733	8	15.51
9	3.325	9	16.92
10	3.940	10	18.31
11	4.575	11	19.68
12	5.226	12	21.03
13	5.892	13	22.36
14	6.571	14	23.68
15	7.261	15	25.00
16	7.962	16	26.30
17	8.672	17	27.59
18	9.390	18	28.87
19	10.117	19	30.14
20	10.851	20	31.41
25	14.611	25	37.65
29	17.708	29	42.56
30	18.493	30	43.77
35	22.465	35	49.80
40	26.509	40	55.76
50	34.764	50	67.50

Degrees of Freedom = (n - 1)

Appendix E
Area Under the Normal Curve

(percentage of area from mean to given Z-score)

Z	.00	.01	.02	.03	.04	.05	.06	.07	.08	.09
0.0	0.00	0.40	0.80	1.20	1.60	1.99	2.39	2.79	3.19	3.59
0.1	3.98	4.38	4.78	5.17	5.57	5.96	6.36	6.75	7.14	7.53
0.2	7.93	8.32	8.71	9.10	9.48	9.87	10.26	10.64	11.03	11.41
0.3	11.79	12.17	12.55	12.93	13.31	13.68	14.06	14.43	14.80	15.17
0.4	15.54	15.91	16.28	16.64	17.00	17.36	17.72	18.08	18.44	18.79
0.5	19.15	19.50	19.85	20.19	20.54	20.88	21.23	21.57	21.90	22.24
0.6	22.57	22.91	23.24	23.57	23.89	24.22	24.54	24.86	25.17	25.49
0.7	25.80	26.11	26.42	26.73	27.04	27.34	27.64	27.94	28.23	28.52
0.8	28.81	29.10	29.39	29.67	29.95	30.23	30.51	30.78	31.06	31.33
0.9	31.59	31.86	32.12	32.38	32.64	32.90	33.15	33.40	33.65	33.89
1.0	34.13	34.38	34.61	34.85	35.08	35.31	35.54	35.77	35.99	36.21
1.1	36.43	36.65	36.86	37.08	37.29	37.49	37.70	37.90	38.10	38.30
1.2	38.49	38.69	38.88	39.07	39.25	39.44	39.62	39.80	39.97	40.15
1.3	40.32	40.49	40.66	40.82	40.99	41.15	41.31	41.47	41.62	41.77
1.4	41.92	42.07	42.22	42.36	42.51	42.65	42.79	42.92	43.06	43.19
1.5	43.42	43.45	43.57	43.70	43.83	43.94	44.06	44.18	44.29	44.41
1.6	44.52	44.63	44.74	44.84	44.95	45.05	45.15	45.25	45.35	45.45
1.7	45.54	45.64	45.73	45.82	45.91	45.99	46.08	46.16	46.25	46.33
1.8	46.41	46.49	46.56	46.64	46.71	46.78	46.86	46.93	46.99	47.06
1.9	47.13	47.19	47.26	47.32	47.38	47.44	47.50	47.56	47.61	47.67
2.0	47.72	47.78	47.83	47.88	47.93	47.98	48.03	48.08	48.12	48.17
2.1	48.21	48.26	48.30	48.34	48.38	48.42	48.46	48.50	48.54	48.57
2.2	48.61	48.64	48.68	48.71	48.75	48.78	48.81	48.84	48.87	48.90
2.3	48.93	48.96	48.98	49.01	49.04	49.06	49.09	49.11	49.13	49.16
2.4	49.18	49.20	49.22	49.25	49.27	49.29	49.31	49.32	49.34	49.36
2.5	49.38	49.40	49.41	49.43	49.45	49.46	49.48	49.49	49.51	49.52
2.6	49.53	49.55	49.56	49.57	49.59	49.60	49.61	49.62	49.63	49.64
2.7	49.65	49.66	49.67	49.68	49.69	49.70	49.71	49.72	49.73	49.74
2.8	49.74	49.75	49.76	49.77	49.77	49.78	49.79	49.79	49.80	49.81
2.9	49.81	49.82	49.82	49.83	49.84	49.84	49.85	49.85	49.86	49.86
3.0	49.87									
3.5	49.98									
4.0	49.997									
5.0	49.99997									

Appendix F
Critical Values for the F-ratio

n_2 \ n_1	1	2	3	4	5	6	7	8	9	10	15	20
1	161	200	216	225	230	234	237	239	241	242	246	248
2	18.51	19.00	19.16	19.25	19.30	19.33	19.36	19.37	19.38	19.39	19.43	19.45
3	10.13	9.55	9.28	9.12	9.01	8.94	8.88	8.84	8.81	8.78	8.70	8.66
4	7.71	6.94	6.59	6.39	6.26	6.16	6.09	6.04	6.00	5.96	5.86	5.80
5	6.61	5.79	5.41	5.19	5.05	4.95	4.88	4.82	4.78	4.74	4.62	4.56
6	5.99	5.14	4.76	4.53	4.39	4.28	4.21	4.15	4.10	4.06	3.94	3.87
7	5.59	4.74	4.35	4.12	3.97	3.87	3.79	3.73	3.68	3.63	3.51	3.44
8	5.32	4.46	4.07	3.84	3.69	3.58	3.50	3.44	3.39	3.34	3.22	3.15
9	5.12	4.26	3.86	3.63	3.48	3.37	3.29	3.23	3.18	3.13	3.01	2.94
10	4.96	4.10	3.71	3.48	3.33	3.22	3.14	3.07	3.02	2.97	2.84	2.77
11	4.84	3.98	3.59	3.36	3.20	3.09	3.01	2.95	2.90	2.86	2.72	2.65
12	4.75	3.88	3.49	3.26	3.11	3.00	2.92	2.85	2.80	2.76	2.62	2.54
13	4.67	3.80	3.41	3.18	3.02	2.92	2.84	2.77	2.72	2.67	2.53	2.46
14	4.60	3.74	3.34	3.11	2.96	2.85	2.77	2.70	2.65	2.60	2.46	2.39
15	4.54	3.68	3.29	3.06	2.90	2.79	2.70	2.64	2.59	2.55	2.40	2.33
16	4.49	3.63	3.24	3.01	2.85	2.74	2.66	2.59	2.54	2.49	2.35	2.28
17	4.45	3.59	3.20	2.96	2.81	2.70	2.62	2.55	2.50	2.45	2.31	2.23
18	4.41	3.55	3.16	2.93	2.77	2.66	2.58	2.51	2.46	2.41	2.27	2.19
19	4.38	3.52	3.13	2.90	2.74	2.63	2.55	2.48	2.43	2.38	2.23	2.16
20	4.35	3.49	3.10	2.87	2.71	2.60	2.52	2.45	2.40	2.35	2.20	2.12
25	4.24	3.38	2.99	2.76	2.60	2.49	2.41	2.34	2.28	2.24	2.09	2.01
30	4.17	3.32	2.92	2.69	2.53	2.42	2.33	2.27	2.21	2.16	2.02	1.93
35	4.12	3.27	2.87	2.64	2.48	2.37	2.28	2.22	2.16	2.11	1.96	1.88
40	4.08	3.23	2.84	2.61	2.45	2.34	2.25	2.18	2.12	2.08	1.92	1.84
50	4.03	3.18	2.79	2.56	2.40	2.29	2.20	2.13	2.07	2.03	1.87	1.78
100	3.94	3.09	2.70	2.46	2.30	2.19	2.10	2.03	1.98	1.93	1.77	1.68

BIBLIOGRAPHY

American Cars Are Getting Better, *Changing Times*, Sept. 83, p. 40

American Telephone and Telegraph, *Statistical Quality Control Handbook*, 11th printing, Delmar Printing Co., Charlotte, NC, 1985

Baker, Edward M., and Artinian, Harry L., The Deming Philosophy of Continuing Improvement in a Service Organization, *Quality Progress*, June 1985, p. 61–69

Barstow, T. S. and Lambert, J. E., Auditing a Quality Program with the Help of Statistics, *Quality Progress*, Milwaukee, November 1982

Besterfield, Dale H., *Quality Control*, Prentice-Hall, Inc. Englewood Cliffs, NJ, 1979

Birnbaum, Mark and Sickman, John, *How to Choose Your Small Business Computer*, Addison-Wesley Publishing Co., 1983

Bronson, Gail, U.S. Firms Plus Foreign Partners Equal Prosperity, *U.S. News and World Report*, March 5, 1984, p. 50–51

Carson, G. B., Bolz, H. A. and Young, H. H., *Production Handbook*, 3rd Ed., The Ronald Press Co., New York, 1972

Clements, Richard, Seeking Q-1, *Manufacturing Success*, Grand Rapids: Solution Specialists, May 1986, (Vol. I No. 4) p. 1

Crosby, Phillip B., *Quality is Free*, McGraw-Hill Book Co., New York, 1979.

Curry, Jess W. and Bonner, David, *How to Find & Buy Good Software*, Prentice Hall Inc., Englewood Cliffs, NJ, 1983

DataMyte Corporation *DataMyte Handbook*, DataMyte Corp., Minnetonka, MN, 1984

Deming, W. Edwards, *Quality, Productivity, and Competitive Position*, M.I.T., Cambridge, Mass., 1982.

Dobyns, Lloyd, *NBC White Paper: If Japan can, Why can't we?* NBC Broadcasting, June 1980. Available on video tape.

Ehrbar, A. F., Grasping the New Unemployment, *Fortune*, May 16, 1983, p. 106–112

Feigenbaum, Armand V., *Total Quality Control*, 3rd Ed., McGraw-Hill Book Co. New York, 1983

Ford Motor Company, *Q-101 Quality System Standard*, Ford Motor Company, Dearborn, Michigan, 1983

Fusfeld, Herbert I. and Haklisch, Carmela S., Cooperative R&D for competitors, *Harvard Business Review*, Nov-Dec 1985, p. 60–76

Gayman, David J., Computers in the Tool Crib, *Manufacturing Engineering*, Dearborn, Michigan, September 1986

Grant, Eugene L. and Leavenworth, Richard S., *Statistical Quality Control*, 5th ed., McGraw-Hill, Inc., New York, 1980

Green Light—The GM/Toyota deal rolls on, *Time*, Jan. 2, 1986, p. 63

Hamper, Ben, I, Rivethead, *Mother Jones*, Sept. 86, p. 26

Hoeffer, El, With SPC, The Quality Goes in before the Parts Come Out, *Purchasing World*, March 1986

Imai, Masaaki, *Kaizen: The Key to Japan's Competitive Success*, Random House Business Div., New York, 1986.

Jackson, Barbara Bund, Build Customer Relationships that Last, *Harvard Business Review*, Nov-Dec 1985, p. 120–128

Jessup, Peter T., *The Value of Continuing Improvement*, Ford Motor Company, 1985, Published in the proceedings of the International Communications Conference, June 1985

Juran, J. M., *Quality Control Handbook*, 3rd ed., McGraw-Hill Book Co., New York, 1974.

Juran, J. M. and Gryna, Frank, *Quality Planning and Analysis*, New York, McGraw-Hill Book Co., 1980

Kukla, Bob, Meeting Customer Needs, *Quality Progress*, Milwaukee, June 1986, p. 15–18

Main, Jeremy, Under the Spell of the Quality Gurus, *Fortune*, August 18, 1986, p. 30

Marks, Mitchell Lee, The Question of Quality Circles, *Psychology Today*, March 86, p. 36

Mason, Robert D., *Statistical Techniques in Business and Economics*, Richard D. Irwin, Inc., Homewood, IL, 1982.

Narus, James A. and Anderson, James A., Turn your industrial distributors into partners, *Harvard Business Review*, March-April 1986, p. 66–71

Orsini, Joyce, "Three Tips for Quality Control", *The Magazine of Bank Administration*, April 1982, pages 45–49.

Perlmutter, Howard V. and Heenan, David A., Cooperate to compete globally, *Harvard Business Review*, March-April 1986, p. 136–152

Ross, Irwin, Corporations Take Aim at Illiteracy, *Fortune*, Sept. 29, 1986, p. 48–54

Powell, Bill, Rebuilding Corporate Empires, *Newsweek*, April 14, 1986, p. 40

Quality Progress, Getting Good Parts: An in-depth look at Supplier Quality Programs, American Society for Quality Control, Milwaukee, November 1984 (entire issue)

Raine, George, Building Cars Japan's Way, *Newsweek*, March 31, 1986, p. 43

Reich, Robert B. and Mankin, Eric D., Joint Ventures with Japan give away our future, *Harvard Business Review*, March-April 1986, p. 78–86

Reid, Robert, Tips for Teaching SPC, *Manufacturing Systems*, August 1986, p. 28–31

Schonberger, Richard J., *Japanese Manufacturing Techniques*, Macmillan Publishing Co., New York, 1982

Schroeder, Michael, If you can't bear 'em, join 'em, *Working Woman*, March 1985, p. 69–71

Taguchi, Genichi and Wu, Yu-in, *Introduction to Off-line Quality Control*, Central Japan Quality Control Association, Tokyo, Japan, 1979

Taguchi, Genichi, *On-line Quality Control during Production*, Japanese Standards Association, Tokyo, Japan, 1981

Taylor, Alexander L., The Growing Gap in Retraining, *Time*, March 28, 1983 p. 50–51

Peters, Thomas J. and Waterman, Robert H., *In Search of Excellence*, Warner Books, Inc., New York, 1982

Young, Robert K. and Veldman, Donald J., *Introductory Statistics for the Behavioral Sciences*, New York, Holt, Reinhart, and Winston, 1972.

The Washington Times, Illiteracy in America, Washington Times, Washington, D.C., Sept. 29, 1986.

Wu, Yu-in, *Quality Engineering: Product and Process Design Optimization*, American Supplier Institute, Romulus, Michigan, 1985